FROZEN IN TIME

FROZEN IN TIME

AN EPIC STORY
OF SURVIVAL AND A
MODERN QUEST FOR
LOST HEROES OF
WORLD WAR II

MITCHELL
ZUCKOFF

HARPER

www.harpercollins.com

HarperCollins books may be purchased for educational, business, or sales pro-
motional use. For information, please e-mail the Special Markets Department
at SPsales@harpercollins.com.

FIRST EDITION

Designed by William Ruoto
Maps designed by Springer Cartographics, LLC

Library of Congress Cataloging-in-Publication Data has been applied for.

ISBN 978-0-06-213343-4 (Hardcover)
ISBN 978-0-06-226937-9 (International Edition)

13 14 15 16 17   OV/RRD   10 9 8 7 6 5 4 3 2 1

For Suzanne, Isabel, and Eve

CONTENTS

A NOTE TO THE READER

THIS BOOK TELLS two true stories, one from the past and one from the present.

The historic story revolves around three American military planes that crashed in Greenland during World War II. First, a C-53 cargo plane slammed into the island's vast ice cap. All five men aboard survived the crash, and their distress calls triggered an urgent search. Next to go down was one of the search planes, a B-17 bomber, stranding nine more men on the ice. Finally, a Coast Guard rescue plane called a Grumman Duck vanished in a storm with three men aboard while trying to save the B-17 crew.

For nearly five months, through the Arctic winter of 1942–1943, survivors and their would-be saviors fought to stay alive and sane in the most hostile environment on earth, clinging to life in snow caves and the tail section of the B-17. As the war raged on, America's military tried to rescue the icebound men by land, sea, and air, sometimes with fatal results. When hope seemed lost, a legendary aviator from the early days of flight devised a half-mad plan to land a seaplane on a glacier.

I learned about these events while hunting through newspaper archives for hidden treasures: stories that once captivated the world, only to fall through the cracks of history. After too many brassiere ads to count, I came across a 1943 series of newspaper articles titled "The Long Wait," about the crew of the wrecked B-17. Intrigued, I dug deeper, collecting declassified documents, maps, photographs, interviews, and previously unknown journals, seeking critical mass for a book.

Along the way, I stumbled upon a loose-knit society of men and women determined to locate the Grumman Duck's resting place and to bring home the remains of the three heroes it carried. Driving that effort was a tireless dreamer named Lou Sapienza, a photographer-turned-explorer who dedicated himself physically, financially, and emotionally to finding the frozen tomb of three men he knew only from faded photographs. Through Lou and his cohorts I met families who'd been waiting nearly seven decades for the return of their lost loved ones.

I also connected with Duck-devoted Coast Guardsmen who believed that all hands should be present and accounted for, one way or another. One in particular, Commander Jim Blow, committed himself with the same fervor he once gave to his work as a search-and-rescue pilot. Soon I realized that I couldn't tell the full story of the three crashes without also writing about the modern mission of the Duck Hunt.

In the summer of 2012 I joined Lou, Jim, and their combined civilian and military teams on a remote glacier in Greenland, where we experienced the world's unfinished attic firsthand. Together we used cutting-edge technology, an overlooked military crash report, and a yellowed treasure map complete with an X to solve one of the last mysteries from World War II.

Although written as a narrative, this is a work of nonfiction. As explained in my note on sources, I took no liberties with facts, dialogue, characters, details, or chronology. Because the story moves between past and present, date markers such as "November 1942" and "October 2011" signal which tale is being told. Also, the historical story is written in past tense, while the modern story is in present tense.

I played a role in the Duck Hunt, and I appear in the book, but it isn't about me. It's about ordinary people thrust by fate or duty into extraordinary circumstances, one group in 1942 and another group seventy years later. Separated by time but connected by character, their bravery, endurance, and sacrifices reveal the power of humanity in inhumane conditions. I hope I've done them justice.

—Mitchell Zuckoff

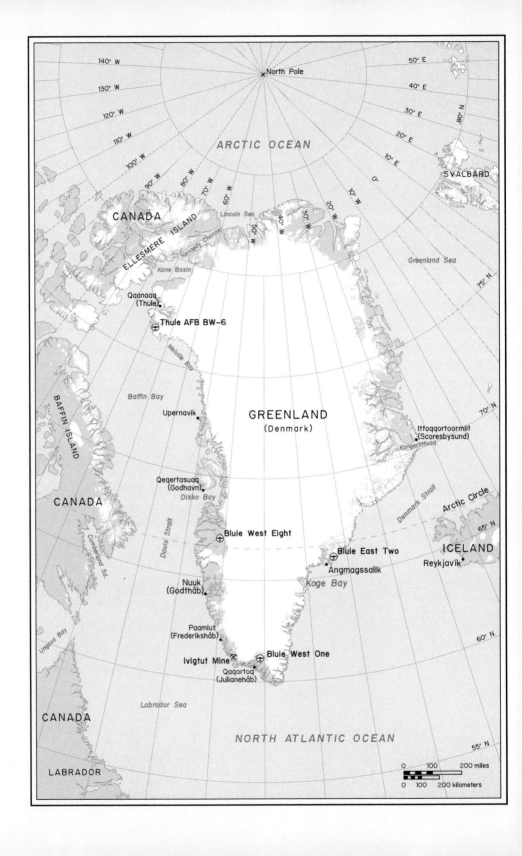

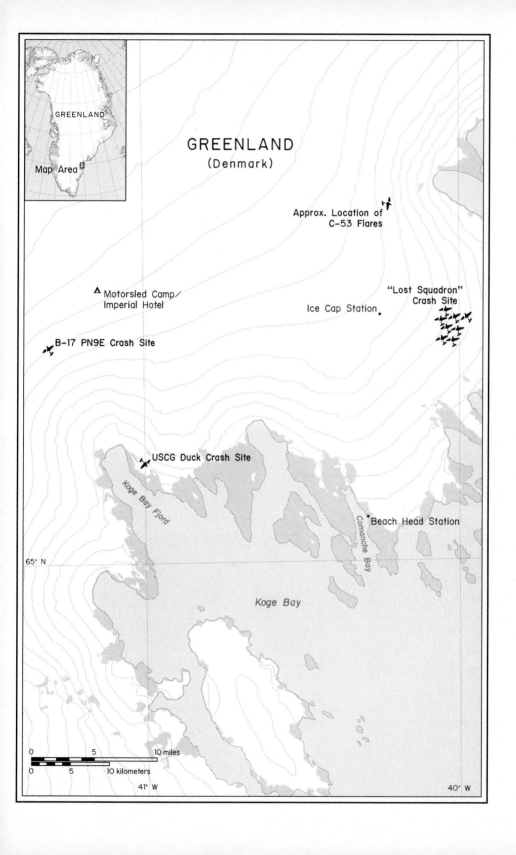

FROZEN IN TIME

PROLOGUE

THE DUCK

NOVEMBER 1942

ON THANKSGIVING DAY 1942, at a secret U.S. Army base on the ice-covered island of Greenland, a telegraph receiver clattered to life: "Situation grave. A very sick man. Hurry." The message came from the crew of a crashed B-17 bomber, nine American airmen whose enemy had become the ruthless Arctic.

Two and a half weeks earlier, while searching for a missing cargo plane, the crew's Flying Fortress had slammed against a glacier in a blinding storm. Since then, the men had huddled in the bomber's broken-off tail section, a cramped cell in a prison of subzero cold, howling winds, and driving snow. Guarding them on all sides were crevasses, deep gashes in the ice that threatened to swallow them whole. Some crevasses were hidden by flimsy ice bridges, each one like a rug thrown over the mouth of a bottomless pit.

One crew member had already fallen through an ice bridge. Another struggled to keep from losing his mind. A third, the "very sick man," watched helplessly as his frozen feet shriveled and turned reddish black, the gruesome signature of flesh-killing gangrene.

Their only hope was that someone would answer their distress calls, tapped out in Morse code over a battered transmitter rebuilt by their young radio operator. His frozen fingers curled in pain each time he hit the telegraph key: "Dot-dot-dot-dot; dot-dot-dash; dot-dash-dot . . ."

"H-u-r-r-y."

THREE DAYS LATER, in the predawn darkness of November 29, 1942, John Pritchard Jr. and Benjamin Bottoms lay sleeping in their bunks aboard the U.S. Coast Guard cutter *Northland*. The ship rocked gently at anchor in an inlet on Greenland's southeastern coast, a place known to the Americans as Comanche Bay. As the United States neared the end of its first year in World War II, the *Northland* and other ships on the Greenland Patrol kept a lookout for German subs, ferried soldiers and civilians to U.S. bases, and watched for icebergs in shipping lanes.

But when the need arose, they set aside routine jobs for their highest calling: rescue work. They risked their lives and their ships to save sailors lost at sea and airmen whose planes had crashed on the huge, largely uncharted island. Other military branches were

THE U.S. COAST GUARD CUTTER *NORTHLAND* DURING WORLD WAR II. ITS AMPHIBIOUS PLANE, THE GRUMMAN DUCK, IS FAR RIGHT. *(U.S. COAST GUARD PHOTOGRAPH.)*

America's swords and spears; the Coast Guard was its shield. John Pritchard and Ben Bottoms embodied that mission as pilot and radioman of the *Northland*'s amphibious plane, a "flying boat" called a Grumman Duck.

The downed cargo plane that had set the search effort in motion remained lost, each day bringing its five crew members closer to death by cold, starvation, or both. But the nine marooned men of the B-17 bomber crew had been spotted, about thirty miles from Comanche Bay as the Duck flies. The question was how to rescue them from a glacier booby-trapped with crevasses. John Pritchard had an answer. His plan had worked once already, and he intended to put it—along with himself, Ben Bottoms, and their Duck—to a far greater test this day.

Pritchard and Bottoms scrambled out of their bunks and into their flight suits, insulated shells that discouraged the cold but couldn't defeat it. After a fortifying breakfast, they climbed into the Duck's tight tandem cockpit. Pritchard, twenty-eight years old, an ambitious lieutenant from California, sat up front at the controls. Bottoms, twenty-nine, a skilled radioman first class from Georgia, sat directly behind him. Fur-lined leather helmets sat snug on their buzz-cut heads. Goggles shielded their eyes. Heavy gloves held their hands.

As they belted themselves in, Pritchard and Bottoms could glance west to see the deceptively beautiful island, a massive white moonscape left over from the last ice age. Beyond a fringe of gray-black rock at the coastline lay hundreds of thousands of square miles of unbroken frost. If they'd had time for reflection, the contrast between Greenland's enormous ice cap and their peculiar little plane might have inspired an adaptation of the Irish fishermen's prayer: "Dear God, be good to me. The sea is so wide and my boat is so small."

If Pritchard or Bottoms had doubts, neither showed it. To the contrary, their *Northland* crewmates thought the Duck's masters appeared eager to get going. In fact, they moved with purpose bordering on urgency; the sun shone for fewer than five hours a day this time of year near the Arctic Circle, and the two Coast Guard airmen hoped to make not one but two round-trips between the ship and the B-17 crew before darkness returned.

Adding to their rush, as well as their risk, the weather was deteriorating. Snow was falling, and a veil of fog was closing in. At eight o'clock in the morning, visibility was a generous twenty miles. Soon it would be less than four miles. By noon, the snowstorm would be in full force and the sky would be a grayish-white shroud. Visibility would be less than one mile and dropping fast.

The Duck hung from heavy ropes suspended over the *Northland*'s deck. The ropes supporting the Duck were attached to a sturdy steel pole called a boom. At Pritchard's signal, the *Northland*'s crew swung that boom out over the ship's side, to lower the plane and its men into the frigid bay. The ropes unspooled, their pulleys rattling with complaint, as the Duck moved foot by foot closer to the greenish water.

Pritchard and Bottoms steadied themselves as the Duck splashed down next to the ship, then bobbed like its feathered namesake. It was 9:15 a.m.

ALTHOUGH EVERYONE ABOARD the *Northland* knew the plane as the Duck, its formal name was the Grumman model J2F-4, serial number V1640. Thirty-four feet long, fourteen feet high, with a wingspan of thirty-nine feet, the Duck was roughly the size of a school bus. It looked as though it had about the same chance of getting airborne. Even pilots who loved it said the Duck handled with all the grace of a milk truck. Its nickname came from its looks and its mallardlike ability to take off and touch down on either water or land. Because it was slow, awkward, and looked like a collection of spare parts, wise guys called it the "Ugly Duckling."

Pritchard and Bottoms's craft had double wings similar to those of a World War I biplane. Mounted under its narrow silver-gray fuselage was a long metal pontoon that resembled an oversize surfboard. To some, the big float looked like the swollen bill of a Disney duck, adding more justification for the nickname. Smaller floats shaped like miniature torpedoes hung under the tips of each lower wing. A nine-cylinder, eight-hundred-horsepower engine spun a single three-blade propeller. The Duck's maximum speed was listed as 192 miles per hour, but pilots called that a joke. Maybe, they said, if it was headed straight down at full throttle.

A GRUMMAN DUCK SIMILAR TO THE ONE FLOWN BY JOHN PRITCHARD AND BENJAMIN BOTTOMS FLIES OVER GREENLAND. *(U.S. COAST GUARD PHOTOGRAPH.)*

Beneath the Duck's cockpit was a cramped compartment in the fuselage designed to fit a few boxes of cargo or two grown men. Pritchard and Bottoms had stripped the space to its bare essentials so they might squeeze in three or possibly four survivors. Before strapping himself into the rear cockpit, Bottoms had placed two hastily built stretcher-sleds in the otherwise empty compartment. The sleds were to be used to haul B-17 crew members too hurt or too weak to hike from the wreck to a relatively flat, crevasse-free stretch of ice more than a mile away, where Pritchard intended to land. The crewman with ice-block feet would certainly need a sled ride to reach the plane. So would the one with a broken arm and frozen toes. Those two men were the priority passengers on this day's rescue plan.

As they prepared for takeoff, Pritchard and Bottoms were buoyed by the knowledge that, although dangerous in untold ways, their mission was possible. The previous day, they'd made a similar round-trip between the ship and the glacier, returning with two less seriously injured men from the B-17. One had frostbitten feet, the other had broken ribs and frostbitten toes, and both were thin and haggard. But now the two rescued airmen were sipping hot coffee, eating all the soup they could stomach, and being pampered in the *Northland*'s sick bay.

The Duck's deliverance from its shipboard nest to the rolling sea happened regularly, but this morning it drew special attention from the 130 officers and enlisted men aboard the *Northland*. Many lined the rail to watch, their breath making vapor clouds in the crisp, salty air. Every man among them knew what the Duck's crew had already accomplished. They also knew where the Duck was headed, and why. Their presence on deck was the way they showed appreciation and paid respect. They wanted to bear witness to the best among them, two men on their way to becoming legends.

WITH THEIR AUDIENCE rapt, Pritchard and Bottoms detached the rope umbilical cords and set the Duck free. Pritchard set the

Duck's propeller to the "climb" position. He adjusted the air-fuel mixture, primed the engine, hit the starter, and ran through the final items on the preflight checklist. He reached to his right to engage the water rudder, then taxied away from the mother ship.

When he reached an open stretch of Comanche Bay to use as a runway, Pritchard disengaged the water rudder and pulled back on the control stick. He pressed his gloved left hand on the round ball atop the throttle lever, moving it smoothly forward. The engine roared a throaty rumble in reply. The Duck gained speed, bouncing from one wave crest to the next across the choppy water, each impact rattling the bones of its pilot and radioman. A fountain of white spray foamed behind the Duck's tail. A V-for-Victory-shaped wake pointed back toward the *Northland*.

Pritchard jostled the control stick forward and back as he searched for the sweet spot, the proper position for takeoff. With each move of the stick, he fought to keep the Duck's nose level as the plane's speed approached fifty miles per hour. Making the task harder was the fact that forward visibility from the pilot's seat in a Duck is almost zero. Pritchard had to perch high in the cockpit and crane his neck to see in front of the plane, or turn his head from one side to the other to gauge whether he risked colliding with a stray bit of iceberg riding low in the water.

About a quarter-mile from the ship, Pritchard increased the Duck's speed to sixty miles an hour, then sixty-five. The stubby little plane rose from the water and took tentative flight. At first, the Duck flew barely a foot above the waves. Pritchard pulled back on the control stick to gain altitude. The Duck answered, rising several hundred feet into the air. Pritchard pointed west toward Greenland and the desperate men waiting in the B-17's tail. The Duck grew smaller in the eyes of the *Northland*'s crew until it disappeared. It was 9:29 a.m.

As they flew off, Pritchard and Bottoms knew that they had

volunteered for fogbound flights, snow-filled skies, perilous landings, and hazardous takeoffs from icy terrain. They accepted the
job without complaint or hesitation, without the promise of fame
or reward. And they did so with every intention of succeeding.

Yet as Coast Guardsmen and rescue fliers, Pritchard and Bottoms couldn't help knowing the wry, unofficial motto of their service: "You have to go out, but you don't have to come back."

For nearly seventy years after the morning of November 29,
1942, the echo of that phrase haunted the Coast Guard and the families of Pritchard, Bottoms, and a third man, a crew member of the
B-17 they'd tried to save. Before more time passed, before memories
faded and the world moved on, an unlikely group of adventurers,
explorers, servicemen and -women, amateur historians, and profes-
sional scientists committed themselves to proving that saying false.

One way or another, they insisted, the Duck and its men *did*
have to come back.

**LIEUTENANT JOHN PRITCHARD JR. (LEFT) AND RADIOMAN FIRST CLASS BENJAMIN
BOTTOMS.** *(U.S. COAST GUARD PHOTOGRAPHS.)*

1

GREENLAND

2000 BC TO AD 1942

GREENLAND MAKES NO sense.

First there's the name, which as most schoolchildren know should be Iceland, but that was already taken. Almost nothing green grows in Greenland, where more than eighty percent of the land is buried under deep ice. Deep, as in, up to ten thousand feet, or two solid miles. If all of Greenland's ice melted—a worst-case scenario of climate change—the world's oceans would rise by twenty feet or more.

Greenland's colorful name is blamed on a colorful Viking called Erik the Red. Erik went to sea when he was exiled from nearby Iceland in the year 982, after he killed two men in a neighborhood dispute. In addition to being an explorer, a fugitive killer, and a lousy neighbor, Erik was the world's first real-estate shill. He christened his discovery Greenland in the belief that a "good name" would encourage his countrymen to settle there with him. The ploy worked, and the community that Erik founded on the island's southwest coast survived for more than four centuries.

Unlike the Pilgrims who came to North America, Erik and his band found no nearby natives to trade with or learn from. So they relied on themselves and on imports from Europe. But by the Middle Ages, decades passed between ships. The once-robust Vikings grew smaller and weaker. Eventually they died out altogether, leaving ruins but little else. Erik the Red is perhaps better remembered for siring Leif Eriksson, who sailed to North America some five hundred years before Columbus. Leif called his discovery Vinland, or Wineland. But Icelanders wouldn't be fooled twice by the same family, and no lasting settlements followed.

A competing but equally odd theory says that the name Greenland was bestowed by the native Inuit people, formerly called Eskimos by outsiders. Their sporadic presence on Greenland traces back some four thousand years, starting with travelers believed to have crossed the narrow straits from North America. The Inuit clustered near the rocky coastline and in the words of one medieval historian, Adam of Bremen, had "lived there long enough to have acquired a greenish tinge from the seawater beside which they dwelt." Under this theory, anyone who looked vaguely green must have come from Greenland.

If Greenland had to be named for a color, white seems the obvious choice. But blue was viable, as well. Although white at the surface, glacier ice on much of Greenland comes in translucent shades of blue, ranging from faint aquamarine and turquoise just below the surface to indigo in the depths of crevasses. The phenomenon is caused by countless years of snow being compacted into ice. Snow contains oxygen, which scatters light across the visible spectrum, making it appear white. Compacting squeezes out the oxygen, and the compacted ice crystals that remain absorb long light waves and reflect short waves. The shortest light waves are violet and blue. And so, the ice at the cold heart of Greenland is blue.

GREENLAND'S STRANGENESS IS compounded by its great but politically inconsequential size; its almost complete emptiness; and its unconscionable weather.

In a world where size generally matters, Greenland's doesn't. The island is globally overlooked despite being enormous: more than sixteen hundred miles from north to south, and eight hundred miles at its widest point. Greenland could swallow Texas and California and still have room for a dessert of New Mexico, Arizona, Florida, Pennsylvania, and all of New England. It's three times the size of France, and it occupies more than twice the area of the planet's second-largest island, New Guinea.

Yet Greenland is the world's loneliest place. With fifty-eight thousand residents, it has the lowest population density of any country or dependent territory. Only Antarctica, with no permanent residents, makes Greenland seem crowded. If Manhattan had the same population density as Greenland, its population would be two.

One way to picture Greenland is to look at a world map and find the blank white spot to the northeast of North America. Another way is to imagine an immense bowl filled with ice. At the outer edge of the island, jagged mountains that rise as high as twelve thousand feet create the bowl's rim. The land between the coastal mountains, the bowl's concave middle, is filled with ice that built up over tens of thousands of years, as yearly snowfall exceeded melting. The more the ice accumulated, the more the land in the central part of the island became depressed from the weight. Hence the ice-filled bowl that is Greenland.

A closer look reveals that the bowl's rim has cracks—spaces between the mountains. Driven by gravity, large bodies of ice called glaciers flow toward the sea like slow-moving rivers. When a glacier's leading edge runs out of land, it fulfills its destiny by hurling itself piece by piece into the water. The process, called calving, is loud and violent and magnificent. Big pieces of glaciers are reborn as icebergs, some big enough to sink an unsinkable ship. In summer 2012, a glacier in northwest Greenland gave birth to an iceberg the size of Boston. The smallest icebergs are known to Coast Guardsmen as "growlers" because they make sounds like snarling animals when trapped air escapes from inside.

Most photographs of Greenland's glaciers and their iceberg offspring fail to capture their grandeur. They look on film like frothy meringue in a cookbook. In reality, they are unstoppable giants that have conquered the world multiple times, and they wouldn't hesitate to unleash a new ice age if given the chance.

Although the bowl-of-ice analogy is useful, it overlooks an important feature of Greenland. Unlike the smooth, rounded rim of a bowl, the coastline is a ragged, sawtooth affair, with innumerable fjords cutting as deep as ninety miles into the land. As a result, Greenland's coast is more than twenty-seven thousand miles long, a distance greater than the circumference of the earth at the equator.

Even more than its size, Greenland's most defining feature is its climate. Temperatures vary along a spectrum of discomfort, ranging from bone-rattling to instant frostbite. In many places, temperatures regularly reach the only place on the thermometer where Celsius and Fahrenheit agree: 40 degrees below zero. To be fair, at the more habitable southern coastline, the average yearly temperature is about 30 degrees Fahrenheit—habitable, but not necessarily reachable by sea. For much of the year the north of Greenland is ringed by solid pack ice, and the waters to the south are beset by the *storis*, a twenty-mile belt of floating icebergs.

Then there's the wind. In fall and winter, devastating blizzards known as *piteraq* storms race more than a hundred miles per hour across the unbroken landscape. The wind blows glacial dust that can scour glass or blind eyes left unprotected. Soldiers stationed at an American base in Greenland during World War II sometimes crawled from one building to the next to avoid violent winds. An officer who stepped blithely out of his hut was thrown twenty feet into a wall, breaking both arms.

ALTHOUGH GREENLAND'S NATURAL defenses discouraged settlement, some hardy souls insisted. In 1721, two centuries after Erik the Red's colony vanished, Europeans returned to Greenland, led by a Danish-Norwegian missionary named Hans Egede. Hoping to discover Viking descendants, Egede instead found Inuit people, so he stayed to spread the gospel. Colonization followed, though few Danes saw the point of the place. Unlike the native North Americans, the native Inuit people of Greenland never surrendered their majority status to outsiders, though they did embrace Christianity.

Soon the Danish monarchy claimed Greenland as its own. But in contrast to typical colonial relationships, the Danes did so with a benign hand and heavy subsidies. They brought the Inuit people food and manufactured goods, accepting in return animal skins, seal oil, and fish. Overall, Denmark kept Greenland isolated. The Danes' benevolent motives were to preserve the traditional Inuit way of life and to protect the natives from diseases against which they had no defenses. The Danes also feared that outsiders would exploit the natives in trade deals; think beads and trinkets for Manhattan. The world respected the Danish rules, mostly because no one saw much value in Greenland. To this day, the island remains politically attached to Denmark, but Greenlanders have begun transitioning to self-rule.

For the most part, then, Greenland passed the millennia as a giant afterthought. The world's attention did turn to the island in 1888, when Norwegian explorer Fridtjof Nansen led a six-man team across the ice cap, the first such crossing in recorded history. In the early part of the twentieth century, the biggest news about Greenland was a 1933 survey flight over the island by Charles and Anne Lindbergh on behalf of Pan American Airlines.

All of that changed on April 9, 1940, when Nazi Germany invaded Denmark. American leaders suddenly looked with fear upon the big island so close to North America. They shuddered at the thought of Hitler building air bases and ports in Greenland, from which they imagined he might strike at Allied planes and ships in the North Atlantic. Even more frightening, Greenland was then six hours by air from New York, well within the range of German bombers. Worst of all was a doomsday scenario under which the island would be used as a Nazi staging area and springboard for a blitzkrieg, or "lightning war," with a ground invasion of the United States and Canada.

More immediately, American officials worried that Germany would establish elaborate weather stations in Greenland. The weather in Europe is "made" in Greenland; winds and currents that flow eastward over the island give birth to storms heading toward Great Britain, Norway, and beyond. Whoever knows today's weather in Greenland knows tomorrow's weather in Europe. Allied planners feared that German weather stations in Greenland could guide Luftwaffe bombing runs over Great Britain and the Continent. The battle to control Greenland wasn't a war for territory, one American official said—it was "a war for weather."

Concern about Greenland also reflected the fact that some wars are lost not in the field but in the factory. If the Nazis ruled Greenland, Germany would gain control of a rare and unique resource that could help determine the outcome of the war. A mine at Greenland's southwestern coast, in a place called Ivigtut, was the world's only reliable natural source of a milky white mineral called cryolite.

Cryolite, a name derived from Greek words meaning "frost

THE CRYOLITE MINE AT IVIGTUT. *(U.S. COAST GUARD PHOTOGRAPH.)*

stone," was essential to the production of aluminum, and aluminum was essential to the production of warplanes. The malleable, lightweight metal sheathed bombers and fighters, cargo haulers and transport planes. Aluminum skins encased Flying Fortresses and Lightnings, Skytrains and Helldivers, Thunderbolts and Ducks.

German factories used a synthetic version of cryolite, but American and Canadian airplane makers relied on the real thing. At less than a mile from the water, the Ivigtut mine was vulnerable to sabotage or attack. A few well-placed shells from a German battleship, or a bomb placed by Nazi saboteurs, would destroy the mine. Without a reliable supply of cryolite, North American airplane factories would go idle at the worst possible time. The official Coast Guard history of World War II puts it bluntly: "Had the Nazis succeeded in preventing the production and shipment of cryolite, they could have dealt a crippling blow to the Allies."

Greenland, ignored for most of human history, suddenly mattered.

WITH GREENLAND LOOMING over North America, President Franklin D. Roosevelt ignored the Nazi invasion of Denmark, diplomatically speaking. The United States continued to recognize the Danish ambassador to Washington as the legitimate represen-

tative of his country and its territories. A deal was reached to send
well-armed former U.S. Coast Guardsmen to Ivigtut to guard the
cryolite mine and prevent possible sabotage by mine workers of
dubious loyalties. Five Coast Guard ships were also dispatched to
the Greenland coast, where they put weapons ashore to defend the
mine. Canada and Britain made their own secret preparations to
safeguard Ivigtut. But those were temporary solutions.

In April 1941, while the United States was still neutral in the
war, it reached an agreement with Denmark's government-in-exile
under which American forces would protect Greenland against
German aggression by building U.S. air bases and military instal-
lations on the island. Germany was none too happy, but the Nazis limited their response to a propaganda campaign that accused
America of plotting "the enslavement, miscegenation and ultimate
extinction of the native population." Ironic, considering the Nazis'
own policy on "ultimate extinction" of certain peoples.

By the summer of 1941 the United States had assembled a
small fleet of Coast Guard ships and converted fishing trawlers
into what it called the Greenland Patrol. A large part of the patrol's job was to help the U.S. Army establish bases for ferrying
planes to Britain and to defend Greenland against German operations. The patrol, which included the cutter *Northland* and its
Grumman Duck biplane, also protected U.S. ships loaded with
food destined for England, to foil Hitler's plan to starve the British
Isles into submission. Among the ships' other jobs was to monitor icebergs in shipping lanes, extending a role the Coast Guard
had performed flawlessly since joining the International Ice Patrol
in 1912, after the sinking of the *Titanic*. Leading the Greenland
Patrol was Rear Admiral Edward "Iceberg" Smith, a veteran commander who had a PhD in oceanography from Harvard.

By the time the United States entered the war in December
1941, its Greenland air bases were well under construction. Greenland was known in U.S. military code as "Bluie," so the most

important island base was called Bluie West One, built on the southwestern coast near where Erik the Red had first settled. Men stationed there adopted an unofficial motto, "Phooey on Bluie," complete with an eleven-stanza poem by the same name. Among the lines: "The mountains of Greenland rise into the sky; at a field on the fjord where the ice drifts by. The worst of it is, with its infinite trials, there isn't a woman for one thousand miles."

Also important was Bluie West Eight, located in remote west-central Greenland some thirty-five miles north of the Arctic Circle. A key base on the southeast coast, still being built, was Bluie East Two, with the upbeat radio code name "Optimist." In time, the military would have fourteen Bluie sites, some little more than heated wooden shacks, nine on the west coast and five on the east.

Upon entering the war after Pearl Harbor, the United States began a massive buildup of the U.S. Eighth Air Force in Britain. During the summer of 1942, the military began Operation Bolero, in which warplanes flew the "Snowball Route" to Britain, hopscotching from Maine to Newfoundland to Greenland to Iceland to Scotland, refueling along the way. Before the winter of 1942–1943, some nine hundred planes would join the war effort via the Snowball Route. Greenland's Bluie bases, built along the route's main street, were transformed from quiet, remote outposts to hectic, remote outposts.

But as soon as American planes began flying over Greenland, they began crashing into Greenland. Three B-17s ditched on the very first day of flights, in June 1942. The next month, in July 1942, a squadron of six P-38 Lightning fighters and two B-17 Flying Fortress bombers encountered bad weather between Greenland and Iceland. They tried to fly under the storm, through it, above it, and around it, with no luck. Next they tried to turn back to Greenland, to land at Bluie West Eight. But their fuel ran low, and they lost their bearings. With no other options, the pilots of "The Lost Squadron," as it became known, crash-landed on the ice cap near Greenland's east coast.

The first plane, a P-38, tried to land with its wheels down. It was a mistake that caused the nose wheel to dig into the surface snow, flipping the fighter onto its back. The rest of the pilots took note and landed wheels up and belly down, treating their war-planes like giant sleds. Incredibly, all twenty-five men aboard the eight planes survived, with only a few minor injuries.

Three days later, American radio operators picked up their distress signals and dispatched cargo planes to drop supplies. Summertime conditions were so good that the men joked about their situation. "Don't send any more toilet paper," they radioed. "Send women." A week later, rescuers reached them by dogsled and led them to the coast. After a seventeen-mile hike, the men of the Lost Squadron were taken aboard the Coast Guard cutter *Northland* to recuperate.

For more than forty years, the eight planes sat abandoned. As snow and ice piled up atop the planes, warplane enthusiasts dreamed of re-covering the Lost Squadron. In 1992, a team found the planes buried under 268 feet of ice. They dug and melted three vertical shafts down to one of the P-38s and used water cannons to carve a two-thousand-square-foot cavern around it. After hauling it to the surface piece by piece, they reassembled it and restored it to flight. They named the rebuilt fighter *Glacier Girl* and sent it touring at air shows.

Other crash landings in Greenland during World War II ended far worse.

A B-17 FLYING FORTRESS BOMBER AND ITS CREW, FROM THE LOST SQUADRON, ON THE GREENLAND ICE CAP. *(U.S. ARMY AIR FORCES PHOTOGRAPH.)*

2

"A MOTHER THAT DEVOURS HER

CHILDREN"

NOVEMBER 1942

ON THE MORNING of November 5, 1942, a pudgy twin-engine plane called a C-53 Skytrooper took flight from an American air base in Reykjavík, Iceland. Five American airmen were aboard, with Captain Homer McDowell Jr. at the controls. McDowell and his crew had been on a routine cargo mission, a "milk run," and now they were headed west, back to the crew's home base on the far side of Greenland.

Their destination was the airfield at Bluie West One, where McDowell intended to touch down on a treacherous landing strip carved between two mountains. As one pilot put it, the base was "fifty-two miles up a fjord with walls several thousand feet high, numerous dead-end offshoots, no room to turn around, and usually an overcast below the tops of the walls. You had to get it right the first time." Yet as wartime assignments went, it was safer than flying over Germany.

The C-53 Skytrooper's name was more impressive than its

looks. Compared with the fighters and bombers crowding the skies, the Skytrooper was like a station wagon in a showroom filled with Jaguars. Modified from the civilian DC-3 passenger plane, the C-53 Skytrooper was a good craft to use when there was a job to do, but not necessarily an urgent or high-profile one.

This day, McDowell's C-53 was one of eighteen American planes making the six-hour trip from Iceland to Greenland's west coast. The preflight briefing was uneventful, though the weather report included a few trouble spots: patches of snow, overcast, with fog along the route.

McDowell, who had celebrated his twenty-ninth birthday four days earlier, flew the C-53 uneventfully during the first two-plus hours of the trip. The weather held clear as the plane passed over the Denmark Strait and reached Greenland's eastern coastline. Because they made it that far without incident, the men aboard the C-53 had the good fortune of being over land, rather than water, when their luck changed.

A C-53 SKYTROOPER DURING WORLD WAR II. *(U.S. ARMY AIR FORCES PHOTOGRAPH.)*

Shortly after the plane reached the southeast coast of Greenland, a location that defined the edge of nowhere, disaster struck: McDowell's Skytrooper went down on the ice cap. By some accounts, the crash occurred when one of the plane's two engines failed, but other reports were silent on why the C-53 experienced what the military called a "forced landing." The official crash report declared the cause "unknown and no reason given in radio contacts." A handwritten notation added, "100 percent undetermined."

Whatever the cause, McDowell and his copilot, Lieutenant William Springer, managed to keep the plane from breaking apart. All five men aboard survived with no major injuries. This was especially impressive because McDowell had logged only seventy-seven total hours as a pilot. Because the plane remained intact, the men had shelter from the elements and a working radio transmitter to send distress signals. That meant a fighting chance for McDowell, Springer, and their crew: Staff Sergeant Eugene Manahan, Corporal William Everett, and Private Thurman Johannessen.

At that moment, down on the ice when they expected to be up in the air, the five men had every right to feel shaken, cold, lost, and frightened. But they also could count their blessings. They had survived with an intact plane, and the U.S. military had a good record when it came to Greenland rescues: the crews of the Lost Squadron had been saved under similar circumstances four months earlier and perhaps no more than ten miles away. McDowell and his crew had reason to believe that they'd soon be back at Bluie West One with hot coffee in their mugs, warm food in their stomachs, and one hell of a tale.

But not all the news was good. The Lost Squadron had gone down during the long, warmer days of summer. McDowell's crew was on the ice at the cusp of winter, with shorter days, stronger winds, and colder, unpredictable temperatures.

They fired up the radio transmitter and tapped out the last four digits of the plane's identification number, 5-5-6-9, then the urgent message: "Down on Ice Cap." They provided would-be rescuers with their last known location while in flight: latitude 61 degrees, 30 minutes north, and longitude 42 degrees, 30 minutes west. That spot placed them over the water, near the southeast coast of Greenland. Assuming that they had continued flying straight toward their destination, that meant the C-53 had gone down somewhere south of the Bluie East Two radio beacon, located at a coastal village called Angmagssalik. They signed off their first message with a simple "All OK."

A half hour later, McDowell's crew sent a second distress message that gave their altitude as ninety-four hundred feet above sea level. At first, that message caused confusion, as rescuers thought it meant that they had slammed into a mountain. Later, however, rescuers realized that it was their airborne altitude when they noted their last known latitude-longitude coordinates. In a subsequent message, the C-53 crew said they believed they had crashed at an altitude of about two thousand feet above sea level.

An American military radio operator in Reykjavík picked up the first messages and tried unsuccessfully to contact the C-53 crew. He relayed the messages to the American Bluie bases in Greenland, and a search mission took shape. Rescue for McDowell and his men seemed imminent.

WHEN MCDOWELL'S C-53 went down, Greenland was bustling with planes en route to the war as part of Operation Bolero, which meant no shortage of potential searchers. When the weather cleared the day after the crash, U.S. Army Air Forces officials detoured six Britain-bound B-25 Mitchell bombers, along with a C-53 Skytrooper and two PBY Catalina flying boats. At least one B-17 bomber also joined the search that day. Each plane was assigned a search area of about forty square miles to scour

Greenland's east coast. The Coast Guard sent a converted fishing trawler, part of the Greenland Patrol, to search along the coastline, as well.

Optimism rose when a search plane made radio contact with McDowell's crew. The pilot explained: "I asked C-53 what his position was, and he sent back that he didn't know. Then I asked if he was on Ice Cap or Coast, to which he sent back 'Coast.'" The search pilot asked the C-53 radio operator to switch to a different radio channel, but received no reply. "There was a lot of interference on the air and a very weak signal only was coming from the lost plane." The northernmost American base, Bluie West Eight, also heard radio messages from McDowell's crew.

Searchers seemed to have more than enough clues to succeed. But during the frustrating days that followed, no sign of the plane turned up. The initial search planes were joined, and in some cases replaced, by as many as eight other B-17 bombers, twenty C-47 Skytrain cargo planes, and fourteen more C-53s. The search area was extended. When that didn't work, it was extended again.

The air-and-sea searches were the primary rescue efforts, but they weren't the only ones. During the first day of the air search, McDowell's crew made radio contact with a Greenland rescue base called Beach Head Station, a small outpost built near the spot where the Coast Guard cutter *Northland* had picked up the rescued crewmen from the Lost Squadron.

McDowell's crew told a radio operator at Beach Head Station that they could see water in the distance. That night, at two agreed-upon times, McDowell and his men fired flares into the sky above them. Lights believed to be the flares were spotted almost due north of Beach Head Station. That meant the flares apparently had been fired from a spot on the ground somewhere near a weather shack called Ice Cap Station, a vacant army outpost only occupied during the summer.

Believing that he had enough information, and enough moxie,

to rescue McDowell's crew, in stepped a remarkable army lieutenant who was the commanding officer of both Beach Head Station
and Ice Cap Station.

Max Demorest was thirty-two and dashing, with thick, wind-
swept hair, a toothy smile, and a strong, aristocratic chin. Married and the father of a young daughter, Demorest was considered
equally brilliant and brave by his friends. He'd first visited Greenland as an undergraduate at the University of Michigan, having
spent a winter there with a professor to establish a meteorological
station. During the decade before the war, he'd earned a doctorate from Princeton University, a research post at Yale University,
a Guggenheim Fellowship, and a job as acting head of the Geology Department at Wesleyan University. Along the way, Demorest won acclaim for discoveries about the movement of glaciers,
achievements that placed him on the verge of becoming one of the

MAX DEMOREST IN GREENLAND. *(U.S. ARMY PHOTOGRAPH.)*

youngest fellows of the Geological Society of America. After Pearl Harbor, the professor who first brought Demorest to Greenland, William S. Carlson, became a colonel in the U.S. Army Air Forces. Eager to join Carlson and the war effort, Demorest left his family and his laboratory, and volunteered for the miserable conditions of a wartime posting in the frozen north.

Although Professor/Colonel Carlson liked and admired Demorest, he worried that his protégé's fearlessness bordered on recklessness. After their first expedition together, Carlson turned their adventures into a book, published in 1940, called *Greenland Lies North*. In it, he described Demorest's tendency to head off on solo expeditions far from their camp. Carlson issued an ominous warning: "I hoped that Max's ignorance of fear would be chastened," he wrote. "If not, Nature in winter Greenland is a mother that devours her own children."

WITH THREE SERGEANTS in tow, Demorest left Beach Head Station and set a course for Ice Cap Station, seventeen difficult miles away. Demorest planned to leave two sergeants at the station to monitor the radio. Then he would head toward the apparent location of McDowell's downed plane with the third sergeant, a newcomer to Greenland named Donald Tetley. No stranger to the outdoors, Tetley was a slim, quiet Texan who'd worked as a ranch hand. Demorest had taught Tetley the ways of subzero survival, and the two had become a team.

Because the radio message from McDowell's C-53 said the crew could see water, and because Demorest knew the odd twists and turns of the Greenland coast, he suspected that the plane was five to ten miles north of Ice Cap Station. He, Tetley, and the two other sergeants set out on the northward journey on two small motorsleds, each one a hybrid of a toboggan and a two-seat motorcycle with rubber snow treads.

The motorsled teams saw the C-53's flares again on November 8, three days after the crash. Excited by the sight, Demorest thought the rescue would take them no more than three or four days if the weather cooperated. It didn't. After leaving the two sergeants at Ice Cap Station, Demorest and Tetley were pummeled by storms. Their sleds were bedeviled by mechanical failures. Three days after starting out, the two would-be rescuers turned back in disappointment. They returned to Beach Head Station for replacement motorsleds. More than a week would pass before they could venture out again.

Each day, the radio signal from the C-53 grew weaker. Planes that were expected to fly longer routes over water, such as big bombers like the B-17 Flying Fortress, were equipped with emergency radios powered by hand cranks in case they had to ditch in the ocean. The C-53 Skytrooper had only its suitcase-sized transmitter and receiver in the radio compartment. The downed plane had no power, so the crew had to rely on dying batteries, with no way to recharge them.

During one radio exchange, searchers asked McDowell's crew to send a continuous signal on a frequency reserved for emergencies. Known as "transmitting MOs," the process would allow a pilot in a search plane to use the direction finder in his aircraft to home in on the magnetic orientation of the signal—its MO. By pinpointing the MO and steering his plane in that direction, a search pilot could follow an invisible path of radio waves to the signal's source. But the plan was a bust; the C-53's weakened batteries wouldn't allow a continuous signal. Instead, McDowell's crew sent messages at set intervals, every half hour at first, to conserve the batteries. Even that wasn't enough. Days after the first messages, the radio on the C-53 went silent.

From the moment they crash-landed, McDowell's crew faced the twin threats of cold and hunger. If they had been on a mail run, they could have broken into care packages laden with treats. Or, if they'd been carrying cargo, there might have been food rations or other supplies destined for a military base. But their cargo bay was empty, and with just a six-hour flight planned from Reykjavík to Bluie West One, the crew carried few meals. Rescuers estimated that they had two days' rations on hand, at most. Heightening the crew's misery, the plane had no gear for long-term survival on the ice—no heavy clothing or sleeping bags, no emergency stoves or lanterns. And without power from the plane, they couldn't generate heat. The temperature inside their Skytrooper ranged from an estimated high of 15 degrees Fahrenheit to a deadly low of minus 10.

In short, surviving the crash had used up their luck.

Despite having provided rescuers with latitude and longitude coordinates; despite the fact that searchers had seen lights believed to be the C-53's flares; despite multiple radio communications; despite dozens of search flights tracing grid patterns in forty-square-mile boxes, McDowell's crew was no closer to being found. The motorsled team of Demorest and Tetley had turned back; none of the airborne searchers had caught a glimpse of the C-53; and the Coast Guard trawler had seen nothing along the coast. Greenland's immensity was showing. The Skytrooper was sixty-four feet long and had a wingspan of ninety-five feet. On the surface of the frozen island, it might as well have been a dust speck on a hockey rink.

One search pilot's daily log registered the futility: "Orders came . . . to conduct a search for the C-53. On the next day, we went out in the assigned area to search for the C-53 again. Flew in the area of Bluie East Two. The air was very rough. At one time, we lost two thousand feet and at another time gained about fifteen hundred feet. Flew three hours, thirty-five minutes. No luck."

Doubts grew about the flare sightings. An official account of the rescue effort noted that "there was little probability of the plane being in this area," but the search was extended in that direction nevertheless. Later, it was suggested that what searchers thought were flares were instead shimmering lights from the aurora borealis. Yet that might have been rationalization, an attempt to explain searchers' failure to find the C-53 and its five men despite radio contacts and apparent flare sightings. Blaming the northern lights also might have been a way to relieve feelings of guilt and dread. The men writing the official logs of the search knew that each fruitless day brought McDowell, Springer, Manahan, Everett, and Johannessen closer to the end.

ADDING TO THE crew's problems, snowstorms made search flights difficult or impossible for several days after the crash. If many more days passed, or if windblown snow covered the C-53, time would begin running out for McDowell and his men. Whatever rations they had would soon be crumbs. The cold would slow their movements and their minds. Their enemy was no longer Hitler and Hirohito, but hypothermia.

When the C-53 crew members' body temperatures fell below 95 degrees Fahrenheit, shivering would become pronounced. As it fell further, their extremities would turn blue. Severe hypothermia would set in when body temperatures fell below 82 degrees. At 68 degrees, doctors would label their condition "profound." Laymen would call it horrible. Speech would become labored or slurred. They'd grow confused. Their hands would become useless. Some men might become agitated or irrational. Some might lose their wits and try burrowing in the plane or the snow in a desperate bid for warmth. Paradoxically, some might strip off their clothes in delusional fits. Some might fall into a stupor. As their bodies stopped trying to stay warm, their heart rates, breathing, and blood pressure would drop. Two breaths a minute. Then one. Their pupils would dilate. Major organs would fail.

At some point during the physical decline, some might succumb to a toxic mix of anxiety and sadness. In the heat of battle, there's no time to think about a telegram from the War Department arriving at a loved one's door. In the cold of a wrecked air-plane on the Greenland ice cap, there's nothing but time.

McDowell's wife, Eugenia, was waiting for him in Riverside, Illinois. Springer's wife was in West Palm Beach, Florida. The three enlisted men were unmarried: Manahan, from Saybrook, Illinois; Everett, from Pasadena, Maryland; and Johannessen, from Alamo, Texas. Their parents would be notified.

The first Western Union telegrams would inform their loved ones that the men were overdue to return. A later round of telegrams would say they were missing. If they remained lost for twelve months, they'd be declared dead.

3

FLYING IN MILK

NOVEMBER 1942

ON THE SAME day that Homer McDowell's C-53 went down on the ice, a fresh-from-the-factory B-17 bomber touched down on the runway at Bluie West One. It still had what pilots call the "new plane smell," a bouquet of solvents that rivals the "new car smell" for its power to create fond and indelible sensory memories.

The Flying Fortress had hopped from Presque Isle, Maine, to the Allied air base at Goose Bay, on Canada's eastern coast, and from there across the Labrador Sea, which separates Canada from Greenland. The big bomber was supposed to continue its eastward journey the following day to Iceland, en route to its final destination, an American airfield in Britain. B-17s were a primary weapon in Allied bombing campaigns against German targets, so new ones were in great demand. Before the war ended, some twelve thousand Flying Fortresses would fill the skies.

But soon after landing in Greenland, this particular B-17 was diverted from its rendezvous with Nazi Germany.

DURING THE SEVEN years since the first prototype rolled off the assembly line, B-17 bombers had undergone major and minor revisions, from changes in the rudders, flaps, and windows, to being lengthened by ten feet, to having a gunner's position added to the tail. The plane cooling its engines on the Bluie West One runway was a B-17F, the latest and most advanced Flying Fortress yet.

The long-range, high-flying bomber was renowned for being able to dish out and take a ferocious amount of punishment, yet still land in one piece. Just over 74 feet long and 19 feet high, it had a wingspan of nearly 104 feet. The B-17F had four engines and room for eight thousand pounds of bombs, almost double the capacity of its E-model predecessor. It flew at up to 325 miles per hour at twenty-five thousand feet, cruised at 160 miles per hour, and had a range of more than 2,000 miles. For

A B-17F FLYING FORTRESS OVER THE ATLANTIC. *(U.S. ARMY AIR FORCES PHOTOGRAPH.)*

protection, it had the heavy armament that spawned its Flying Fortress nickname, with eleven .50-caliber machine guns. It deserved its almost mythic reputation as a bird of war.

On bombing runs, a crew of up to ten men would include five machine gunners. A bombardier would sit in the plane's cone-shaped Plexiglas nose for a bird's-eye view of potential targets. There, he'd operate the highly classified Norden Bombsight, a computerlike device that guided the delivery of destruction. About a foot high and sixteen inches long, resembling a compact telescope, Norden Bombsights were supposedly able to place a bomb within a hundred-foot circle when dropped from a plane flying at twenty thousand feet. Bombardiers boasted that the device could guide a bomb into a pickle barrel. In fact, the bombsight's accuracy and its secret weapon status were overstated. Nevertheless, the Norden Bombsight was considered so crucial to the war that American bombardiers took a special oath:

> Mindful of the fact that I am to become guardian of one of my country's most priceless military assets, the American bombsight . . . I do here, in the presence of Almighty God, swear by the Bombardier's Code of Honor, to keep inviolate the secrecy of any and all confidential information revealed to me, and further to uphold the honor and integrity of the Army Air Forces, if need be, with my life itself.

The bomber that landed at Bluie West One on November 5, 1942, had a new Norden Bombsight, even though it had yet to be assigned a bombardier. It had machine guns, but didn't yet have machine gunners. It didn't have a nickname—fierce, like *Cyanide for Hitler*, or glamorous, like *Smokey Liz*, or goofy, like *Big Barn Smell*. And it didn't yet have a curvaceous Vargas girl painted below the pilot's window. All that would come with its permanent crew.

For the moment, the new and unpedigreed bomber was the ward of the Air Transport Command, a military shuttle service whose job was to ferry planes to U.S. and overseas bases. Until its combat crew came on board, the untested bomber would be known by its serial number, 42-5088, or more often by its prosaic radio call sign, PN9E.

AFTER THE RADIO distress calls from McDowell's C-53, the temporary crew of the B-17 PN9E got word that instead of going to England, they'd remain in Greenland and join the search for the missing cargo plane. The war would wait, but freezing American airmen wouldn't.

The B-17's pilot, a low-key lieutenant from California named Armand Monteverde, spread word of the new assignment among his ferrying crew: Lieutenant Harry Spencer, the copilot; Lieutenant William "Bill" O'Hara, the navigator; Private Paul Spina, the engineer; Private Alexander "Al" Tucciarone, the assistant engineer; and Corporal Loren "Lolly" Howarth, the radio operator.

This was their first foreign mission after several cushy months of delivering planes around the United States, mostly to bases in the Midwest. To celebrate their maiden overseas trip, the crew posed for a photo outside the plane and had their copies signed by commanding officers in the Air Transport Command. Painted on the plane's side, above where the men stood, was a fitting slogan for a bomber: "Do unto others before they do unto you. . . ." The photo had a serious purpose, too; some ferrying crews crashed in the ocean en route to Europe. A friend of one PN9E crew member joked about that risk, telling him, "Goodbye, sea food."

Also aboard the PN9E was Private Clarence Wedel, a thirty-five-year-old mechanic from Canton, Kansas. Wedel, whose name rhymed with "needle," had hitched a ride in Goose Bay, on his way to a posting in Scotland. Now that the PN9E would be searching for the missing C-53, Wedel would be an extra pair of eyes.

On November 6, the first day of the hunt for McDowell's lost plane, the crew of the PN9E saw nothing but the unbroken white canvas of the ice cap. To Paul Spina, the engineer, it was a beautiful sight. "For miles and miles, all you could see was a level sheet of ice with not one object to blot out its whiteness," the twenty-six-year-old native of upstate New York wrote in his journal. "Along the edges were glaciers floating down to the sea to form icebergs. Later we learned to call them 'iceberg factories.'" After several hours, the PN9E returned to Bluie West One, where the crew learned that other search planes had no luck, either.

The next day, foul weather grounded all the search planes. The day after that, rescue planners assigned the PN9E to a newly

THE ORIGINAL SIX-MEMBER CREW OF THE PN9E: (BACK ROW, FROM LEFT) NAVIGATOR WILLIAM "BILL" O'HARA, PILOT ARMAND MONTEVERDE, AND COPILOT HARRY SPENCER; (FRONT ROW, FROM LEFT) ASSISTANT ENGINEER ALEXANDER "AL" TUCCIARONE, RA-DIOMAN LOREN "LOLLY" HOWARTH, AND ENGINEER PAUL SPINA. (COURTESY OF PETER TUCCIARONE.)

mapped search box that included the spot where the C-53's flares appeared to have been fired. The area was defined by a jagged stretch of east-west coastline with three inlets, or fjords, carved by glaciers into Greenland's bedrock. The area was known as Koge Bugt, or Koge Bay, after a bay near Copenhagen where the Danes had drubbed the Swedes in a 1677 naval battle. The Danish pronunciation stumped the American airmen, so they made it sound like a town in the Midwest, pronouncing it "koh-gee."

Koge Bay is a big bite out of the Greenland coastline, some thirty miles across and fifty-five miles long. Several rocky islands dot the bay, including one called Jens Munks O, a miniature Greenland complete with its own little ice cap. Glaciers pour like lemmings into the waters of the bay, filling it with enormous, sculptured icebergs. Each of the three fjords in Koge Bay had its own name, but the largest and most westerly of the three was called simply the Koge Bay fjord. Native Greenlanders called it Pikiutdlek, or "place where, when we first arrived, there was a bird's nest." In Greenland, birds are uncommon enough to merit special note.

On November 8, three days after McDowell's C-53 went down, the PN9E took flight toward Koge Bay, but bad weather at low altitudes made it impossible to see the ice cap below. Less than two hours into the flight, the B-17's number-four engine lost oil pressure, so pilot Armand Monteverde and copilot Harry Spencer turned back to Bluie West One for repairs and to spend the night. When they landed, they learned that they'd get another chance: McDowell's plane was still missing.

THE NEXT DAY, November 9, 1942, the PN9E drew the same assigned search area. As the crew warmed the engines and prepared for takeoff, two men walked over and introduced themselves as Tech Sergeant Alfred "Clint" Best and Staff Sergeant Lloyd "Woody" Puryear. Best and Puryear worked in the communications department at the base, and they had the day off. They were friends with several men aboard McDowell's C-53, and they wanted to volunteer as searchers. Best also confessed that he wanted to experience a Flying Fortress firsthand. Monteverde welcomed them aboard, and the two volunteers squeezed into the B-17's transparent nose to serve as forward spotters.

With the six-man ferrying crew, plus Wedel, Best, and Puryear, the PN9E was ready for another search flight over the ice cap. As the nine men sat in the plane awaiting clearance for takeoff, they learned that another radio message had been received from the C-53, but it was too faint to comprehend or for rescuers to lock on to its position. The men aboard the PN9E understood: the cargo plane's batteries were nearly dead, and the crew might soon be, too.

The six original PN9E crew members made a bet among themselves: whoever spotted the lost plane would win dinner and drinks, paid for by the other five when they reached England. Wedel, Best, and Puryear were left out because they weren't expected to be around long enough to collect.

As THEY TAXIED the plane for takeoff, Monteverde and Spencer received a radio call from base operations telling them to pull off the runway. Another search plane needed to land because of engine trouble. Spencer was a friend of the other plane's pilot, so he razzed him over the radio. The pilot shot back that he welcomed the abuse: while Spencer would be flying in the cold, he'd be tucked into a "nice, warm sack." Spencer laughed it off, but that phrase would stick in the PN9E crew's collective memory.

The PN9E took flight and headed east across the frozen island. The spotters called out whenever they saw something black against the ice. But each time, on a second or third pass, they'd realize it was an outcropping of rock. About two hours into the flight, the big bomber reached the edge of Koge Bay. They approached from the sea, but again the area was beset by lousy weather. A blizzard blew snow across the surface ice, and swirling winds tossed the B-17 like a rowboat in the ocean. Paul Spina, the engineer, went into the cockpit to ask Monteverde, whom he called "Lieutenant Monty," why they weren't turning back. Monteverde told him they planned to do so as soon as he could find a hole in the weather.

Spina went to the radio room behind the cockpit and told radioman Loren "Lolly" Howarth to call Bluie West One to say they were returning. Bill O'Hara, the navigator, was sitting in the radio room, smoking a cigarette. Also crammed into the small compartment were assistant engineer Al Tucciarone and passenger/spotter Clarence Wedel. With nowhere left to sit, Spina removed his flight jacket and flying boots, bundled them into a pillow, and lay down on the floor.

Monteverde and Spencer tried to get weather reports from Ice Cap Station and Beach Head Station. They couldn't reach either outpost by radio, so they were on their own.

Their assigned search box included an area of the ice cap that stretched about thirty miles north of Koge Bay. They headed in that direction, hoping that they'd be able to circle around the weather and find clear skies. The top layer of bad weather was at about seven thousand feet. They could have tried to fly above it, but their job was to search for a downed plane with five men in peril. That meant keeping their altitude as low as possible while they were in their search box.

Trouble arrived quickly. The route that Monteverde chose to escape the storm instead steered them into a cruel trap of nature. When they reached the end of Koge Bay fjord, Monteverde and Spencer looked through the windshield of the PN9E and saw that everything outside was the same frightening shade of whitish gray. They couldn't tell where the sky ended and the ice cap began.

Harry Spencer thought he saw a horizontal line of blue sky in the distance, and he hoped that they could fly under the overcast to reach it. But the blue line vanished; it was an illusion, a false horizon, created by reflections cast by ice crystals whipped through the sky by an approaching storm.

When the true horizon disappears in the Arctic haze, a pilot might as well be blind. Pilots fortunate enough to survive the phenomenon describe the experience as "flying in milk." It's so common in Greenland that the effect even happens on the ground. Once on a hazy day, Monteverde straightened up too fast after bending over at the waist. Surrounded by whiteness, with no way to distinguish between earth and sky, Monteverde felt as though he were floating inside a giant cotton ball. He lost his balance and fell over backward, laughing at the absurdity of it. But it wasn't funny in the pilot's seat of a bomber with eight other men aboard.

Adding to their plight, Monteverde and Spencer couldn't trust their instruments. The B-17's altimeter measured the plane's altitude above sea level, not above ground level. If the ground beneath their wings rose sharply, as it often did near the Greenland coast, the altimeter would be no help.

Monteverde and Spencer knew they had to act fast. One option would be to turn the PN9E back toward the water. But they were in the airmen's equivalent of a polar bear's den: any movement might wake the beast. With no idea of their altitude, the B-17 might be only a few feet above the ice cap. If Monteverde banked too hard to make the turn, he might dip the wing far enough to make contact with the ground, destroying their plane and putting them in mortal danger. Another option was to pull back hard on the control stick to gain altitude, but that wasn't much better. The big bomber would need time for that, and there was no telling how much room they had dead ahead—the glacier might rise faster than a B-17 could. A third option, the least attractive, would be to continue ahead and hope for the best, risking a nose-first rendezvous with the ice cap.

Monteverde and Spencer faced the classic definition of a dilemma: a wrenching choice among several lousy options—turn, climb, or do nothing. Monteverde gripped the control wheel and made his choice.

4

THE DUCK HUNTER

OCTOBER 2011

WALKING DOUBLE-TIME THROUGH baggage claim at Reagan National Airport in Washington, Lou Sapienza is frazzled. Normally upbeat, Lou frisks himself like a man who's misplaced a winning lottery ticket. He can't find the address of a building in Alexandria, Virginia, where we're supposed to meet a team of military and government officials who Lou believes hold the key to a glorious quest.

As the president of an exploration company devoted to recovering lost military aircraft and fallen servicemen, Lou is here to propose a navigational feat about a million times more difficult than finding a suburban office building. That is, locating a small airplane and three men entombed in ice for seven decades somewhere in Greenland. His inability to find the meeting address doesn't inspire confidence.

Lou and I met three months ago, after I learned from an acquaintance that we had a shared interest in three American military planes lost in Greenland during World War II. When we sat down to talk, I told Lou that I was writing about the past, but I couldn't finish the story without following his planned expedition.

"If I'm gonna work with a writer," Lou said then, "why shouldn't I work with the guy who wrote that great book about mountain climbing—what's his name? Jon Krakauer?"

"You're right," I said. "You should work with Krakauer. *Into Thin Air* is a great book. But Krakauer isn't here. I'm here."

Lou smiled, and that was that.

Today I'm supposed to be along for the ride, a silent observer recording what happens, good or bad, when Lou requests government funding. And yet I'm determined to go with Lou to Greenland, and I'd hate to see the hopes of three heroes' families, Lou's dream, and my book plans collapse because of a missed meeting. I tear a sheet of paper from my notebook and give Lou the address.

We hustle outside the airport to hail a cab. As we approach our destination, I realize that Lou's waiting for me to pay the fare. I reach into my wallet and wonder if I should have found him Krakauer's phone number.

WE'RE USHERED INTO a nondescript conference room at the U.S. Department of Defense Prisoner of War/Missing Personnel Office, or DPMO. Established in 1993, DPMO's job is "to limit the loss or capture of Americans who are serving abroad, and to bring home those who are captured or killed while serving our country."

For someone seeking help, guidance, and financial support to find a missing World War II plane and three lost airmen, a visit to DPMO is mandatory. Lou has yet to bring home anyone killed or missing in action, but he's been describing his plans to folks at DPMO and similar agencies for years by phone, e-mail, and personal contacts. He's finally earned a full-blown meeting, to pitch what he has in mind.

Waiting for us is a cast worthy of a television drama series: two army lieutenant colonels, two government historians, a forensic anthropologist, and three Coast Guard officers. One of the Coasties is a commander, one is a lieutenant commander, and the third is a senior chief petty officer. Call the TV show *CSI: MIA*. President Obama's photo watches over the table from a far wall, and the American flag and a black-and-white POW/MIA banner stand nearby.

The meeting starts immediately. Lou seems nervous, caffeinated, or both. He's fifty-nine years old and six feet tall, a large man with a big presence. He has no interest in sports yet carries himself with the rolling gait of a former athlete. He has thick features that suit him, lively blue-gray eyes behind metal-rim glasses, and wavy silver hair that's long enough to curl onto the top of his collar. Lou's voice is loud, with an adenoidal New Jersey accent. He's

LOU SAPIENZA IN GREENLAND IN 2010. *(COURTESY OF LOU SAPIENZA.)*

garrulous and affable, his default posture somewhere between un-daunted and windmill-tilting.

Seated at the head of an oblong table, Lou looks ill at ease in a blue blazer and a red tie. Still, surrounded by uniforms and suits, he's glad that he decided against his first choice of meeting-wear: a khaki explorer's shirt, complete with epaulets, last in style when Stanley went to Africa. "I thought it might be a bit much," he whispered to me earlier.

Lou thanks everyone for coming and passes out folders with DVDs, photos, maps, and documents about his New York–based exploration company, North South Polar Inc., and his nonprofit organization, the Fallen American Veterans Foundation. The fold-ers include colorful embroidered patches four inches in diameter that celebrate Lou's dreamed-of mission. Woven with green, gold, blue, and white thread, the patches feature a sketch of the Grum-man Duck in profile, the dates 1942 and 2012, and a red X on a map of Greenland. The military men at the table, their uniforms bristling with ribbons and stripes for battles already fought, mis-sions already accomplished, seem unlikely to affix Lou's patches onto their sleeves.

Lou explains that he is seeking DPMO's blessing and financial support to find Grumman Duck serial number V1640, excavate it from thirty to fifty feet of ice, and bring home whatever remains of the plane and its occupants. While he's at it, Lou says, he'd like support to find a second plane also presumed to be under Green-land's ice. That plane, a C-53 cargo carrier with five men aboard, went down twenty-four days before the Duck, a crash that indi-rectly set the Duck on its fateful path. Also, although the chances seem slim, Lou wouldn't rule out finding a third plane: a B-17 bomber that crashed while searching for the C-53 cargo plane.

Lou doesn't realize it, but it's clear to me that his audience needs a scorecard to keep track. The simplest way would be to explain the crashes chronologically and in relation to one another. Something like this: Lou hopes to find a C-53 cargo plane that crashed on November 5, 1942, with five men aboard; a B-17 bomber, which was sent to search for the missing cargo plane, that crashed four days later with nine men aboard; and, most of all, a Grumman Duck that crashed on November 29, 1942, with three men aboard, one of whom was a crewman on the B-17. Lou also might have explained that ten American servicemen remain unaccounted for: five from the C-53; the two-man crew of the Duck; two members of the B-17 crew, one of whom was aboard the Duck when it crashed; and one from a failed rescue mission by motorsled.

Deepening the confusion, Lou wanders to other ideas and dreams, ignoring questions from skeptics around the table. As Lou's presentation meanders onward, his audience grows distracted. Several exchange furtive glances, while others keep their heads down and shuffle through the information packages. Lou sallies forth.

Leading the meeting for DPMO is Lieutenant Colonel James McDonough of the U.S. Army; he is soft-spoken but all business, his tiny bristles of hair at full attention, his muscular frame snug in a desert camouflage uniform. McDonough repeatedly tries to steer Lou back on track and to lower his expectations.

"I applaud your effort," McDonough says. But money is limited, the POW/MIA caseload is daunting, and the goal is to clear cases— that is, to find bodies—quickly, efficiently, and economically. It's evident that McDonough considers Lou's plan to be the opposite: slow, inefficient, and expensive.

McDonough is being realistic out of necessity. More than eighty-three thousand U.S. military personnel are "unaccounted for," an overwhelming majority from World War II. Most will never be found—particularly those lost at sea—so the government prefers to shoot fish in barrels rather than chase wild geese, or Ducks, as the case may be.

McDonough tells Lou, "The ones that are easiest to crack usually take priority over the others." One reason is that time is of the essence. DPMO tries to repatriate the remains of missing soldiers, seamen, and airmen while immediate family members are still alive to see them laid to rest.

As a wall clock loudly ticks, hovering over the discussion is the question of money. With military budgets stretched by wars and slimmed by spending cuts, the phrase du jour in the MIA world is "ratio of cost to recoveries." Because of complicated logistics and difficult climate and terrain in Greenland, Lou's unfunded budget exceeds $1 million. McDonough says that might cover thirty searches of European farms and forests, where plenty of World War II–era soldiers and airmen remain unaccounted for. In other words, the price tag alone makes the odds against Lou at least thirty-to-one with this crowd.

Next in line to burst Lou's bubble is Lieutenant Colonel Patrick Christian, who's been patient while explaining intricacies of the government and military bureaucracy. He addresses Lou as he might a child: "Finding a little plane in this big, wide world is not an easy thing," he says. "We want to help. We just don't want to say anything to make you think we can do more than we can."

My translation: You haven't a prayer of getting a dime here. Lou theoretically could mount an expedition on his own, without DPMO's money or formal approval. But he'd have to raise the funds from private sources and win permits from officials in Greenland. The odds of that are slim to none.

Lou plows ahead as though McDonough and Christian have just offered him an all-expenses-paid, fly-drive trip to the Greenland ice cap. Dejected, I doodle in my notebook, sketching Lou's Duck mission patch with a slash through it.

McDonough asks Lou point-blank, "Do you have the resources to accomplish what you want in Greenland without the support of the United States government?"

Lou hesitates before answering: "That's a tricky question." Translation: No.

McDonough clearly doesn't think the question was tricky, but he's game. He asks slower and with more volume, "Can . . . you . . . do . . . this . . . without . . . government . . . support?"

I feel embarrassed for Lou, but he dodges again: "Well, there are all different kinds of support." For starters, he says, it would be great if the government would supply a military C-130 Hercules transport plane to carry his team and equipment to Greenland. That might be worth several hundred thousand dollars.

I begin to wonder if Lou has a hearing problem. McDonough looks down at the table and says nothing. The room goes silent.

After an awkward minute, seemingly out of nowhere, the Coast Guard flies to Lou's rescue. A calm, deep voice from the far side of the table chimes in: "I anticipate that if we get a request for a C-130, we will submit the request." It's the first upbeat note after a two-hour dirge.

The voice belongs to Commander Jim Blow of the Coast Guard's Office of Aviation Forces, who's worked with Lou off and on for several years. Recruiting-poster handsome, trim and fit, with a square chin and short dark hair flecked with gray, Blow is one of many Coast Guard fliers who are the spiritual heirs of two of the service's greatest heroes: John Pritchard Jr. and Benjamin Bottoms, the Duck's missing crewmen.

As Blow knows, of the eighty-three thousand American servicemen and -women who remain unaccounted for, only three served in the Coast Guard. One is a lieutenant who died in a Japanese prison camp during World War II and whose remains are considered unrecoverable. The other two are Pritchard and Bottoms. By helping Lou to bring home the Duck's crew, Blow would be honoring a promise to do everything possible to leave no man behind.

When he agrees to seek government assets on Lou's behalf, or more accurately on behalf of the lost men and their families, Blow sounds tempted to call out the Coast Guard's motto: "Semper Paratus," Latin for "Always Ready."

"I think we can do it," Blow says confidently.

Lou relaxes. We shake hands with everyone around the table and leave.

OVER A STEAK lunch at the airport Chili's restaurant, a cheerful Lou says he's more confident than ever. He's still woefully short on money, and he's nowhere near certain where to dig through the ice for the lost plane. Nevertheless, Lou believes that he and his team, which now includes me, will solve an enduring mystery of World War II: What happened to the Duck and the three men it carried?

As we're leaving, Lou slaps a meaty hand on my shoulder. "That went well, man," he says. "Didn't you think?"

I agree, not mentioning the yawning gap in the budget, the logistical and technical challenges ahead, or the doubts that I shared with my notebook before the Coast Guard arrived, cavalry-style.

A serene look crosses Lou's face. He repeats a comment I've heard him make several times before. Previously, he said it with a question mark hanging almost imperceptibly in the air. This time there's none of that: "We're gonna bring these men home."

The Duck Hunt is on.

5

A SHALLOW TURN

NOVEMBER 1942

THREE HOURS INTO the search, pilot Armand Monteverde was lost in the blinding whiteness that enveloped his B-17. He didn't know his altitude above the ice, what lay ahead, or what was left or right beyond the tips of the PN9E's outstretched wings. Maybe it was a harmless snow-filled cloud. Or maybe it was a concrete-hard glacier. Somewhere below was the downed C-53 that Monteverde and his crew were searching for, but they couldn't see anything in the sky, much less on the ground. More to the point, the missing cargo plane was now a distant second in priority, behind their own survival.

Monteverde was twenty-seven, an unmarried first lieutenant from Anaheim, California. Built like a wrestler, short, stout, and broad-shouldered, he had sad green-gray eyes, a full lower lip, a narrow face, and a Roman nose. Instead of stereotypical pilot bravado, he had a mild manner and a gentle voice that gave him an air of quiet competence. His crewmates ribbed him for being a California boy, but they liked and trusted him. Unlike some officers, they knew that Monteverde had come up the hard way. He'd paid for flight school by working nights in a gas station, then flew for a cargo airline in Mexico before joining the Army Air Forces. A capable pilot, Monteverde had logged seven hundred hours of flight time, though only fifteen of them in the cockpit of a B-17.

His copilot, Harry Spencer Jr., was a twenty-two-year-old second lieutenant from Dallas. Six-foot-one, blond, hazel-eyed, lean, and cowboy handsome, Spencer had a cleft chin and dimpled cheeks. He looked as though he'd been born to wear a white silk pilot's scarf. But he was no arrogant golden boy. Spencer's humble, even-keeled nature made him more suited to service as a ferry pilot than a fighter jock. Smart, well-read, and sensitive to the feelings of others, Spencer was an Eagle Scout who possessed a leader's natural understanding of how to build a team. He'd had a busy

LIEUTENANT ARMAND MONTEVERDE, PILOT OF THE B-17 PN9E.
(U.S. COAST GUARD PHOTOGRAPH.)

year: he married his college girlfriend in April, learned to fly at Southern Methodist University, and joined the Army Air Corps in September.

As they sat side by side in the cockpit, Monteverde and Spencer knew that if their bomber continued on its current heading, every minute of flight time would bring them about three miles farther into the unknown. Only a fool would stay the course, and neither man was a fool. They crossed "maintain present heading" off their mental checklists.

In theory, pulling back on the control wheel and gaining altitude was a possible way out. But at the head of the Koge Bay fjord, the ice cap rises steeply to an estimated eight thousand feet above sea level. The pilots didn't know their position above that

LIEUTENANT HARRY SPENCER, COPILOT OF THE B-17 PN9E.
(U.S. COAST GUARD PHOTOGRAPH.)

rapid upslope, or whether they could put enough separation between their bomber and the ice before time and space ran out. But they did know that gaining that much altitude would take more time than they suspected they had. That eliminated option two, gain altitude.

They had parachutes aboard, but bailing out wasn't a serious option; the plane wasn't on fire or under enemy attack, and the inside of a bomber was more attractive than the outside of the ice cap. Subtly altering course to the east or west might work, but they didn't know whether a new heading a few degrees in either direction would mean a new lease on life or a fatal error. That would be like taking another card in blackjack without knowing what cards they'd already been dealt.

Their last alternative, turning back toward the fjord and the open sea beyond, seemed the best choice in a bad situation. The primary risk would be if, unknown to the pilots, the B-17 was less than one hundred feet above the ground. A world of white is a confusing place, and instruments that gauged altitude above sea level were no help. Pilots in Greenland told stories of flying along blissfully, only to realize they had landed, belly down on the ice. Others sheepishly described preparing to touch down when in fact they remained high above the runway.

If, as Monteverde and Spencer believed, the PN9E's clearance over the ice cap exceeded one hundred feet, they'd have enough room to dip one of the wings and turn the B-17 seaward. If the clearance was less, search planes from Bluie West One would soon be looking for whatever remained of them and their brand-new bomber.

Monteverde and Spencer felt confident that they had enough altitude to execute a turn, so they trusted their guts. Spencer thought they might have as much as one thousand feet of clearance. It was their decision to make, but if they'd polled their crew, no man would have objected to the pilots' logic.

Among those eager to endorse any change of course was Al Tucciarone, a dark-haired twenty-eight-year-old with a proud nose and a winning smile. He'd been a laborer and truck driver back home in the Bronx, but war had transformed him into an assistant engineer on the big bomber.

A week earlier, before leaving Presque Isle, Maine, Tucciarone had sent his fiancée, Angelina, a postcard that read, "Everything is running smoothly. Do not worry. I'm feeling fine. Will see you soon." Now, however, none of that was true: nothing was running smoothly; he was worried; he wasn't feeling fine; and there was a distinct possibility that he wouldn't see Angelina soon. Tucciarone knew that they were in trouble when he looked out a window and couldn't see five feet beyond the bomber.

PRIVATE ALEXANDER "AL" TUCCIARONE, ASSISTANT ENGINEER ON THE PN9E.
(U.S. ARMY PHOTOGRAPH.)

To be careful, Monteverde decided to bank the plane gently at first. He eased the PN9E into a shallow turn.

THE LEFT WINGTIP slapped the ice. The fifteen-ton bomber shook furiously. A terrible crunching sound exploded inside the plane. Men not strapped into their seats flew like confetti. After touching the ground, the wingtip bounced back up, leveling the plane. But the PN9E was no longer flying, and it wasn't a fortress. It was a giant bobsled, sliding, careening, and carving a groove into the glacier. As it slowed, the bomber turned like a weathervane, its nose pointing due north into the wind.

The B-17 skidded more than two hundred yards across the ice, then came to an abrupt halt. The sudden stop catapulted flight engineer Paul Spina through a window on the roof of the radio compartment. Spina landed prone in the snow with no jacket, no shoes, no gloves, and no flight helmet. He'd taken off his jacket and boots when he lay down, and the force of being launched from the plane must have torn off his gloves and helmet. Spina was stunned, bleeding, and exposed to freezing temperatures. Both bones in his right forearm were broken close to the wrist. His hands and feet were cut, and frostbite clamped down on his toes and fingers.

Spina couldn't see the plane or any of his crewmates through the blinding, windblown snow. Partly for warmth and partly in despair, he tried to hide his face in his frozen hands. Another crewman heard him cry out, "Somebody pull me in—I'm freezing." He began to stand, but as he did everything went black. Spina passed out. The dark-haired private, five foot four and less than 150 pounds, would soon freeze to death unless someone helped him.

Also needing help was one of the volunteer searchers, Alfred "Clint" Best. When the plane stopped sliding, Best flew from the bombardier's seat through the broken Plexiglas nose: the PN9E had sneezed him onto Greenland. Twenty-five years old, thickset, quiet, and introverted, a bookkeeper in civilian life, Best suffered a cut on top of his head and a bruised knee. The other volunteer searcher, Best's friend Lloyd "Woody" Puryear, climbed out through the broken nose to pull Best back inside. He suffered cuts and bruises.

Clarence Wedel, who'd come aboard as a passenger in Goose Bay, bounced from one end of the B-17's cabin to the other. Wedel rose from the deck with cuts on his face and a black eye that left his eyeball red and inflamed.

Al Tucciarone, the assistant engineer, and Loren "Lolly" Howarth, the radio operator, were strapped into the radio room's bucket seats, so they fared better. A blow to the chest left Tucciarone weak, with a couple of broken ribs but not grievously injured. Howarth sustained a cut on his head. The three officers, Monteverde, Spencer, and O'Hara, emerged from the crash dazed but unhurt.

Incredibly, all nine men were alive. The plane was another story.

When the PN9E's fuselage struck the ground, the metal buckled and twisted. The fearsome symbol of American air power broke in two, like a balsa-wood model in the hands of an angry child. The break came behind the wings, separating the front section—the nose, the cockpit, the navigator's compartment, and the radio compartment—from the waist section and tail. During construction, a metal band had been riveted into place, almost like a zipper, attaching prefabricated sections of the plane. On impact with the ice, it unzipped. Yet even after breaking apart, both front and rear sections of the bomber had plowed the ice along the same path, as though connected by memory. When the broken B-17 stopped skidding, the nose and tail sections were separated by about a dozen feet, like a salami with a chunk sliced from the middle.

The twelve-foot metal propellers of both left-side engines were shredded. The tips of the right-side propellers curled like ribbons. The metal skin of the fuselage outside the radio compartment was torn away. The left outboard engine hung limp from its mountings. High-octane fuel spilled from the left wing and from auxiliary tanks, drenching the radio compartment and the bomb bay. The PN9E was a wreck.

Still in the cockpit, Monteverde gathered his wits. The collision between the left wingtip and the ice cap came as such a surprise that his mind refused to absorb it. He'd dipped the wing a few degrees, for a few seconds, just beginning his turn. A complete circle was 360 degrees; the PN9E had banked only about 10 degrees when everything went haywire. His immediate, irrational thought was that one of the four engines had flamed out. But Monteverde's head cleared and he understood the truth: the wingtip had sliced into a glacier, and the PN9E was down.

Monteverde heard an ominous hissing sound around the cabin. Convinced that his wrecked plane was on fire, he unbuckled and scrambled out through a broken cockpit window. Once outside, Monteverde realized that the hissing sound was dry, sandy snow thrashing against the metal. He looked around and saw Spina, bloodied and unconscious near the silent left engines. Bill O'Hara followed Monteverde through the window. The navigator leaped into deep snow, soaking his leather boots, and joined Monteverde at Spina's side. The two officers carried Spina inside the B-17's torn-open tail section to treat his wounds.

Soon all nine of the PN9E's crewmen were crammed together inside the bomber's rear end, stunned and freezing. Outside, the snowstorm raged.

ONE CONSOLATION WAS that Monteverde's decision to ease the plane into a shallow turn had probably saved their lives. Had he continued flying ahead, they would have struck the ice cap nose-first, with potentially explosive results. Had they parachuted out, the cold would have killed them if the jump hadn't. If Monteverde had banked hard, pointing the left wingtip sharply downward, the PN9E might have cartwheeled tail over nose when it touched the ground, tearing apart the plane, with predictable results for the men inside.

Later, the military would declare that the crash was caused by "lack of depth perception due to blending of overcast and heavy blowing snow." It was a formal way of saying "flying in milk." In the military way of things, Monteverde received sixty percent of the blame, while the weather was faulted for the remaining forty percent. "The pilot is considered to be responsible for this accident," the official investigation found, "in that he flew over the Ice Cap under an overcast contrary to instructions" during preflight briefings. It continued: "He was overzealous in attempting a hazardous operation and did not have the proper training to accomplish the mission safely." The review board recommended that in the future, pilots like Monteverde "not be sent over this route classified as experienced with the small amount of time shown by this Officer."

He never publicly objected to the finding, but to a large extent Monteverde was blamed for forces beyond his control. He'd been found guilty of being inexperienced, for following orders to conduct a search in bad weather, and for failing to transform himself from a ferry pilot on his first overseas mission into a grizzled Arctic search pilot familiar with the treachery of lost horizons.

For now, though, worrying about blame took a back seat to survival. Monteverde and his crew had been sent over the ice cap to find a crashed C-53. Instead, for the second time in four days, an American military plane had gone down in an undetermined location on the frozen, largely uncharted east coast of Greenland. When the top brass at Bluie West One awoke that morning, five American airmen had been in danger of freezing or starving to death. Now the number was fourteen.

MONTEVERDE AND HIS crew didn't know it yet, but the PN9E had come to rest about seven miles north of the Koge Bay fjord, on a glacier approximately four thousand feet above sea level. On a clear day, the landscape looked from the sky like an unbroken sheet of ice. But up close, it was scarred by windblown waves of snow called *sastrugi* and crisscrossed by deep crevasses. Many crevasses were covered by natural bridges of accumulating snow and ice that made them impossible to see and therefore doubly dangerous. Some ice bridges were strong enough to bear a man's weight. Some weren't.

By luck or momentum, both parts of the broken bomber had somehow glided over a long stretch of the crevasse field. Now, at rest, the bomber's tail sat motionless near the edge of a crevasse that split the ice to an unknown depth. If the crevasse widened, or if the PN9E's rear end slid backward, all nine men who'd taken refuge inside would fall with it into the chasm.

A more immediate worry was the cold. They had no heat, no light, no stove. They had no sleeping bags, no heavy clothing, no Arctic survival gear. A few seconds outside would coat a man's face with frost. In minutes, blood would rush from his extremities to his core. Exposed skin would die. In the sky, the men on a B-17 were warriors. On the ground, they were frozen sardines in a busted-open can.

They had no way to call for help: the crash had badly damaged the radio. The crew was afraid to even try it, fearing that sparks would ignite the spilled fuel they could smell all around them. Unlike the C-53 crew, the men of the PN9E couldn't enjoy morale-boosting, potentially lifesaving contact with the outside world. Unless, that is, radioman Lolly Howarth could repair the smashed equipment, piece together a jury-rigged transmitter, or find an emergency transmitter buried in the wreckage. His crewmates weren't counting on it. They stacked the broken radio gear at the open end of the tail section. The equipment didn't work, so the heavy black boxes could at least act as a windbreak.

When engineer Paul Spina regained his senses in the tail section, he found his crewmates crowded around him, rubbing his frozen hands and feet. Monteverde located a first aid kit and put his Boy Scout training to use. For a half hour, he pulled and twisted on the engineer's arm so the broken bones would line up and knit together. Spina tolerated the rough treatment without complaint or painkillers, and Monteverde admired the small man's toughness. For a splint, Monteverde used a piece of aluminum that Spencer tore from the interior of the ruined plane. Even wrapped in parachute cloth, it felt cold against Spina's skin, but it kept his arm straight. Then Monteverde went man to man, tending wounds with the first aid kit.

As Monteverde ministered to them, crewmen gathered supplies and counted rations. Hoping to make an insulated nest on the floor of the tail section, they arranged seat cushions, blankets, window covers, travel bags, and unfurled parachutes. They found a heavy canvas tarp, normally used to cover the plane's nose between flights, and draped it over the open end of their quarters. They were shivering and wet, but at least the wind and snow wouldn't roam unchecked through the PN9E's tail. Spina marveled that the tarp was in the plane to begin with; it was supposed to have been stripped off before takeoff and left behind at the base. Without it, he thought, Greenland would have made quick work of them.

Tenuous connections to the outside world raised the crash survivors' spirits. Woody Puryear expected to be on the B-17 for a few hours at most, to search for his missing friends in the C-53. Now he swelled with pride when he pulled a silk parachute from its pack and saw the words "Made in Lexington, Kentucky." Puryear was a strapping, twenty-five-year-old country boy, more than six feet tall and 210 pounds. Before the war, he'd worked as a meat cutter and electric lineman in his hometown of Campbellsville,

Kentucky. Lexington was the big city, some sixty miles away. But in a shattered B-17 on Greenland's ice cap, Lexington was a link to family and friends. The parachute label made Puryear pensive. "Memories of home," he'd say, "are best when you're far away—when you don't know whether you'll ever get home again."

The men knew they'd soon be painfully thirsty and hungry. Monteverde blamed the shock of the crash for making them all parched. But all the liquids on board were frozen. Best and Puryear had brought along a thermos filled with hot coffee, but now they opened it to find a block of brown ice. With no way to melt it, the crew ate dry snow. It kept them hydrated, but it made their throats scratchy and wouldn't quench their thirst, no matter how much they swallowed. Spina's hands were too frozen for him to feed himself, so the others filled his mouth with snow.

SERGEANT LLOYD "WOODY" PURYEAR, VOLUNTEER SEARCHER ABOARD THE PN9E.
(COURTESY OF JEAN SPINA.)

Darkness came early in November, so the nine men settled in for the night in their rounded metal cell. When whole, the B-17 stretched seventy-four feet. The torn-open rear section was about half that, and much of the interior space was unusable. The bomber's curved walls, with ribs made of aluminum alloy, narrowed increasingly the closer a man got to the tail. The floor consisted of catwalks normally used by the waist and tail gunners to move through the B-17's rear, or aft, section. Now, the catwalks were the only level places on which to lie down. That meant the nine-man crew of the PN9E had to squeeze onto a platform about fifteen feet long and three feet wide, or about five square feet per man. They tumbled on each other like a litter of puppies, some pressed against the plane's cold, hard ribs.

They wrapped themselves in blankets of cut-up parachute cloth and everything soft they could salvage. They wriggled their toes to keep them from freezing. With each breath they inhaled fumes of splattered fuel. Men wanted to smoke cigarettes, but Monteverde forbade it, fearing that they'd explode their quarters. Spina's friends pressed themselves against him on both sides for body heat. Stretching their legs had to be done in turn. Moving through the scrum was almost impossible, so to change position they grabbed onto the metal butts of the .50-caliber machine guns and used them like subway handholds. When night fell, the pitch-black shelter rang with calls of men trying to avoid stepping on each other: "Is that all right? Am I missing you all right?"

Before they sought the relief of sleep, Monteverde made a modest announcement: "This is as new to me as it is to you. According to regulations, I am in charge. But I want any and all suggestions you might happen to think of. We will work it out together."

AFTER A FITFUL night, they awoke the next morning, Tuesday, November 10, 1942, and rearranged their den for greater warmth and comfort. They found canvas wing covers and added them to the tarp curtain at the open end of the tail section, but the effort to make the compartment weather-tight was futile. Cold wind and fine snow shot through every crack between the tarps. Even the seams of their clothing weren't tight enough to block the sting of wind-driven snow.

The day was too stormy to leave the plane and investigate their surroundings, but several men scrounged around the wreckage. Inside the crushed radio compartment, they stumbled upon the most valuable item of all: an emergency radio transmitter. Waterproof, weighing thirty-five pounds, and painted bright yellow, the radio came with a metal-frame box kite and a reel with eight hundred feet of antenna wire. Unlike the crew of the C-53, which didn't have an emergency radio on board, the men aboard the PN9E wouldn't need to rely on their plane's dying batteries. Power for the transmitter came from a hand-cranked generator built into the housing. The radio had curved sides that allowed a seated man to hold it between his thighs while turning the power crank. The idea was that an airman whose plane ditched in the ocean would sit in a life raft and crank out rescue calls. The radio's hourglass design spawned its affectionate nickname, the "Gibson Girl," after the curvy women in drawings by fashion artist Charles Gibson. One problem was that a Gibson Girl spoke but didn't listen; the radio was a transmitter but not a receiver. Still, a lost man with a Gibson Girl between his legs had a fighting chance at survival.

The wind was too strong to fly the antenna kite their first full day on the ice. But in the following days, Lolly Howarth, the radio operator, flew it whenever the storm died down. Though unsure whether the radio transmitter worked, or whether anyone received its message, Howarth sent steady streams of SOS signals at the universal maritime distress frequency of 500 kilohertz. Serious and quiet, a twenty-three-year-old aspiring actor from Wausaukee, Wisconsin, Howarth soon began worrying that the Gibson

Girl wouldn't save them, after all. He eyed the plane's damaged radio equipment and wondered if he could fix it. The sooner, the better.

An inventory revealed enough K rations—canned meats, biscuits, cereal bars, gum, and other staples—for one man to survive thirty-six days. That meant four days' worth of meals for the nine of them. Monteverde intended to stretch the rations for ten days, knowing that even that might not be long enough.

Each box of K rations included a four-pack of cigarettes. But as much as the tobacco might have relieved hunger, Monteverde continued to ban it for fear of igniting the leaked fuel. They also found six boxes of U.S. Army Field Ration D, the military's code name for chocolate bars. The D rations were included in three "jungle kits" given to the officers: Monteverde, Spencer, and O'Hara. One quirk was that D-ration chocolate sacrificed flavor for heat resis-

CORPORAL LOREN "LOLLY" HOWARTH, RADIO OPERATOR ABOARD THE PN9E.
(U.S. COAST GUARD PHOTOGRAPH.)

tance, so the bars wouldn't melt in soldiers' packs. That was the least of the PN9E crew's worries.

It might seem like backward military logic to give jungle supplies to men flying the Snowball Route over Greenland. But no one complained: the kits also contained long bolo knives that could chop snow and ice.

THE NEXT DAY, November 11, brought no respite from driving snow and subzero temperatures. Cold in Greenland is almost a living thing, a tormenting force that robs strapping men of strength, denies them rest, and refuses them comfort. In time, it kills like a python, squeezing life from its victims.

Again the crew hunkered down. They savored the reduced rations Monteverde distributed and ate as much dry snow as they could swallow. The big treat of the day was a few squares of chocolate. A cycle developed in which their hands and feet froze and then thawed, each time triggering a burning, aching sensation. When it happened, they'd say their extremities had "stoved up." Navigator Bill O'Hara suffered the most from its effects, particularly in his feet.

That night, desperate for a smoke, Paul Spina ignored the spilled fuel and Monteverde's orders. The two had developed a warm rapport during their ferrying duties, and Spina knew that Monteverde was no disciplinarian. When everyone else fell asleep, he unwrapped his bandaged hands with his mouth. Awkwardly using both frostbitten hands, Spina fished out a cigarette and matches from his pocket, stuck the cigarette between his lips, and tucked a matchbox under his chin. When he struck a flame, his crewmates startled awake. Spina calmly lit his cigarette and asked if anyone wanted a drag. The plane didn't explode and neither did Monteverde, who had a soft spot for the affable engineer. Spina also had a bond with copilot Harry Spencer, who soon held cigarettes to his mouth and slipped him extra bits of chocolate.

Now that they could light matches, they used small fires to melt snow in a thermos cup. Little by little they had water to drink, not eat.

During their first three days on the ice cap, as the storm blew itself out, the crew of the PN9E listened for planes overhead, fantasized about long furloughs after being rescued, and got to know one another.

THE PN9E'S FAILURE to return to Bluie West One doubled the job confronting searchers. No one knew where the B-17 had crashed, but the area near Koge Bay that had been assigned to Monteverde's crew for the C-53 search seemed a logical place to look. Despite the storms, on November 10 seventeen planes left Bluie West One to search for the PN9E.

PRIVATE PAUL SPINA, ENGINEER ABOARD THE PN9E.
(COURTESY OF JEAN SPINA.)

Meanwhile, sixteen C-47s and six B-17s went out looking for McDowell's C-53. The skies over Greenland were teeming with search planes diverted from the war. All returned that night with no sign of either missing crew.

The following day, two search flights went out for each of the downed planes, but heavy storms near Bluie West One drove them back to base. The same Arctic weather that contributed to or caused both crashes now conspired to prevent McDowell's C-53 and Monteverde's B-17 from being found.

ALTHOUGH THE SKIES on the west coast of Greenland were stormy, the weather on the east coast gave the men of the PN9E a break on Thursday, November 12. Dawn arrived clear and bright. The crew was weak and tired, but the blue sky gave them a lift. Radioman Lolly Howarth flew the Gibson Girl kite and looked more closely at the damaged equipment from the radio compartment.

Crew members strong enough to work crawled out of their hideout to rake several feet of windblown snow that had piled around and atop the olive-colored plane. The temperature was 30 degrees below zero, but the task kept their minds busy and their bodies active. They hoped that removing the snow would keep the B-17 visible from the sky if a search plane flew overhead. Despite all they'd been through, their spirits remained strong. They told each other that just as they'd been out searching for the C-53, someone would be out looking for them.

As his men kept busy, Monteverde ducked inside the tail section to keep Spina company. They talked awhile, and soon the pilot and the engineer realized that they needed spiritual help. They knelt together to pray.

Meanwhile, copilot Harry Spencer and navigator Bill O'Hara decided to have a look around. Despite O'Hara's frozen feet, he wanted to tough it out. He was twenty-four, the hard-nosed son of a coal mine manager from outside Scranton, Pennsylvania. After working in the mines during his teenage years, O'Hara graduated from the University of Scranton with a degree in business administration. Awaiting him at home was a beautiful girlfriend, Joan Fennie.

Spencer and O'Hara knew that Koge Bay was southeast of their wreck. On clear days like this one, they could see the water. Distances were difficult to calculate across the featureless expanse of ice, but they felt confident the bay was no more than ten miles away. If they could reach it on foot, they might be able to establish their position with greater precision. Maybe they could use the Gibson Girl to hail one of the Coast Guard ships patrolling nearby. The emergency radio had an automatic mode to send SOS signals and also a manual mode for custom messages. Spencer con-

LIEUTENANT WILLIAM "BILL" O'HARA, NAVIGATOR ABOARD THE PN9E.
(COURTESY OF JEAN SPINA.)

sidered using a life raft from the PN9E to paddle along the coast to the weather shack at Beach Head Station.

Even if they couldn't hike to the bay, Spencer and O'Hara intended to plot the locations of nearby crevasses, to keep everyone safe when they ventured away from the plane. They hoped that a map of ice fissures would also provide a pathway to the crash site for rescuers on foot, motorsleds, or dogsleds.

Aware that some crevasses were covered by snow or ice bridges, the two lieutenants walked slowly, testing the ground in front of them before each step. They found one crevasse and made their way around it, then found another and again took evasive action. About fifty yards from the plane, Spencer stepped on what felt like a patch of solid ice.

A moment later, he disappeared.

6

MAN DOWN

NOVEMBER 1942

HARRY SPENCER THOUGHT he was a goner.

As a boy, he'd devoured books about adventures in the Arctic. He marveled at its wonders and respected its dangers. The young Texan knew that he'd fallen through an ice bridge covering a hidden crevasse. He also knew that being swallowed by a crevasse is like being swallowed by a whale: after a brief, exhilarating rush, it rarely ends well.

As he fell, Spencer retained enough presence of mind to understand that his life expectancy was pitifully brief. If the crevasse was deep, say three to four hundred feet, he could expect to live for about five seconds before staining the ice red at the bottom.

A five-second fall might be just enough time for the handsome young lieutenant to see the life that he was supposed to live flash before his eyes. There might be time to picture his pretty young wife, Patsy, and to imagine the pleasures and sorrows they'd share. If Spencer could think as fast as he fell, five seconds might be enough time to envision three fine children, three wonderful grandchildren, and an unbreakable Friday date night on their houseboat. He might see himself endure heartbreaking loss and celebrate great joys; become an admired business, community, and church leader in Irving, Texas; and win local office and an armful of civic honors. He might have time to imagine his own obituary. Yet five seconds wouldn't be enough time to appreciate such an obituary's closing lines: "Harry's intelligence, wit, laughter, sense of adventure, commitment to enabling the city of Irving to be a home for all people, commitment to education and good medical care for all, deep love for the Creator, and devotion to family and friends, touched everyone he met along the way. He believed the Boy Scout motto and lived out the creed in his daily walk. He will be missed!"

With every second, with every foot Spencer fell deeper into the abyss, he moved further away from the rich, full life he seemed destined to enjoy.

Nature wasn't likely to offer any help. Crevasses are caused when adjacent parts of a glacier move at different speeds toward the edge of Greenland. Some portions of glaciers creep while others race, as though eager to fling themselves into new lives as icebergs. As physical tensions rise between fast- and slow-moving portions of a glacier, the ice fractures. The rifts that result are crevasses. Because most crevasses have unbroken vertical walls, Spencer had every reason to believe that he had fallen into nature's equivalent of a twenty- or thirty-story elevator shaft.

But something unexpected and extraordinary happened. After falling through the ice bridge, Harry Spencer didn't drop for five seconds to the bottom of the crevasse. Time seemed to slow as he fell—he felt as though he were falling forever—but in fact he fell for less than three seconds. And that made all the difference.

Spencer's fall was cut short by a block of ice the size of a Jeep that had somehow wedged itself between the walls of the crevasse, below the spot where the ice bridge broke. The ice block created a natural platform large enough to halt Spencer's fall about one hundred feet from the surface. Spencer landed on his back, dazed but unhurt. An agnostic might call it an astonishing piece of good fortune, but Spencer was a churchgoing man, so he'd forever consider it divine intervention.

As he took stock of his situation, Spencer realized that he was covered by a blanket of snow; the top layer from the surface had followed him into the crevasse. He brushed himself off and counted his blessings. He was alive and well, but his predicament wasn't over. Now he had to get out.

Peering downward beyond the edge of the ice platform, Spencer saw that the crevasse continued far below into deep, blue-ice darkness. As he shifted his weight to turn around, the block slipped, threatening to take him crashing to the bottom with it. But the gaps in the walls of a crevasse tend to narrow the deeper they go, so the block wedged itself into the walls again, at a spot just below its original position. Spencer got to his feet to look around. As he

A CREVASSE NEAR THE PN9E CRASH SITE, DATED 1942. HARRY SPENCER IS BELIEVED TO BE ONE OF THE THREE MEN PICTURED. (COURTESY OF CAROL SUE SPENCER PODRAZA.)

gazed left and right, he noticed that his platform was the only one like it as far as he could see. If he'd fallen through the ice bridge a few feet in either direction, he'd almost certainly be dead at the bottom of the crevasse.

Spencer felt strangely serene. He drank in the ethereal beauty of his surroundings, a cathedral of ice in shades of blue and white. His eyes followed the translucent light-blue crevasse walls upward to the white hole through which he fell. He looked up to a sky that was its own winking shade of blue. Sunlit clouds completed a scene fit for a ceiling fresco.

Halfway down the gullet of a crevasse was the last place the twenty-two-year-old newlywed airman expected to find himself. Climbing out on his own wasn't possible; the crevasse walls were too vertical and too slick, and Spencer had no tools to make handholds. But at that moment, encased in natural splendor, standing on his little platform, Harry Spencer felt confident that it wouldn't be his final resting place. He thought: "God must have a plan for me, or this block of ice wouldn't be here."

WHEN SPENCER FELL, Bill O'Hara called for help from the men in and around the carcass of the PN9E. When no one responded, O'Hara repeated the distress call three more times. When he got their attention, O'Hara ordered, "Get rope! Get rope!"

Roused from his prayer session with Spina inside the bomber's tail, Monteverde ran toward the hole, ignoring the risk of crevasses and the frostbite gnawing at both his feet. Following close behind were several others who'd been searching the snow for rations and equipment lost in the crash.

The break in the ice bridge revealed the crevasse to be fifteen feet wide, more than capable of swallowing them all. Fearful of joining Spencer, they tiptoed toward the opening like soldiers approaching a minefield. The PN9E crew had no previous idea what a bridge-covered crevasse looked like, and until the crash none thought he'd ever need to know. Soon they would begin to recognize what Tucciarone described as a telltale sign of danger: "Small ridges, two or three inches in height, drifted over with snow." But not all hidden crevasses had those markings, so every step on the glacier carried deadly risk.

Monteverde dropped to his belly and wriggled to the edge. He shouted down to Spencer, who called back that he was OK.

Other crew members hurried back to the bomber and gathered nylon lines from the parachutes they'd sliced into bedding. They braided six lines into a long, strong rope, followed their footsteps back to the crevasse, and lowered the rope to their copilot. Spencer tied the last few feet into a loop and slipped his head and arms through. The makeshift lasso settled securely under his armpits.

Seven of Spencer's crewmates, all but the injured Spina, grabbed onto the other end of the rope and hauled him toward the surface. With each pull, the nylon threads sawed a narrow channel deeper and deeper into the edge of the crevasse. When Spencer neared the top, Monteverde and the others realized that they were pulling him toward the underside of an impassable shelf of overhanging ice. Already weak and tired, they lowered Spencer back down a hundred feet to the platform. They pulled up the rope and plotted a new strategy.

First, they attached a parachute harness to one end of the rope. On the initial rescue attempt, Spencer felt as though the nylon loop under his arms was cutting him in half. The harness had straps that went over his shoulders, across his chest, and between his legs, and was far more comfortable and secure. Along with the parachute harness they lowered a bolo knife from the survival kit. When everything was in place, the seven able-bodied crewmen returned to the rope line and pulled Spencer back up. This time when he reached the ice ledge near the top, Spencer used the bolo knife to hack from below while Monteverde used another knife to chop away from above. Together they carved a V-shaped notch at the lip of the crevasse, a passage large enough for the rope team to pull Spencer through.

More than three hours after he fell, Spencer was out of the crevasse and back on solid ice. He was in relatively good shape—cold to the bone and suffering from a lost glove and frostbite on his exposed hand, but otherwise unhurt. He might yet enjoy a long and fruitful life, capped by lavish tributes in a distant obituary.

SPENCER'S FALL FORCED the PN9E crew to accept that they'd crash-landed on a glacier shot through with hidden crevasses. Even a cursory inspection revealed that they were hemmed in on all sides. The most worrisome crevasse was the one that crossed underneath the PN9E's tail. Small at first, it grew at an alarming rate. Within days, it would threaten the crew and their living quarters.

If the PN9E had crash-landed in late summer, the ice bridges might have already melted, revealing the crevasses like highways on a road map. But in November they were no more visible at the surface than subway tunnels.

The crevasses forced the PN9E survivors to abandon hope that they could hike to Koge Bay. No longer could they dream of hunting seals on the coast until being spotted by a Coast Guard ship. Any thought of paddling a life raft to Beach Head Station was abandoned, too. They were trapped on the ice cap, their survival dependent on someone spotting them from the air. Yet even if they were found, a new question arose: How would anyone reach them? A plane couldn't land among the crevasses, and motorsled or dog teams would face the same hidden dangers that nearly killed Harry Spencer.

The exhausted, bone-chilled crew of the PN9E trudged back to the broken bomber. They collected scraps of felt from the ruined radio compartment, drained fuel from the plane's tanks, and made a fire on the icy ground. To celebrate Spencer's survival and to replenish their energy, Monteverde issued the men full meal rations. They thawed snow for water to drink and heated canned meat over the open flame. They warmed themselves before the fire and pulled off their boots to rub blood back into their feet.

Then, all nine men prayed together, offering thanks for the survival of their friend, crewmate, and companion. Afterward, they prayed as a congregation every day. There were no atheists in their ice hole.

EVEN BEFORE SPENCER'S rescue, several men were hobbling on frostbitten feet; now they were in worse shape. Men whose shoes or flying boots were leather, as opposed to rubber, suffered the most. Woody Puryear discovered that his unlined leather boots were warm and soft during the day, but every night they froze, encasing his feet in blocks of ice. O'Hara's leather dress shoes hadn't dried since he'd jumped into the snow to help Paul Spina after the crash. Hours of trying to rescue Spencer from the crevasse left O'Hara with no feeling below the ankles.

Inside the bomber's tail, O'Hara confided to Monteverde that he thought his feet were frozen solid. Spina heard the navigator say, "I don't even know if I have any feet or not." When Monteverde helped to remove O'Hara's shoes, the men saw an awful sight: the skin on his feet had deep, ugly cracks, and they'd turned sickening shades of blue, yellow, and green.

Monteverde was stunned to find that O'Hara's feet felt nothing like flesh and bone. An awful comparison rushed to his mind: they felt like the cold, hard metal on the butt ends of the plane's machine guns. Monteverde knew that the cause wasn't just O'Hara's leather shoes. The navigator had been among the toughest and most selfless of them all. His feet had frozen after the crash when helping Spina, and he'd made matters worse by trying to hike with Spencer to the sea. Working to rescue Spencer from the crevasse had been the final straw.

Hoping to reverse or at least limit the damage, Monteverde rubbed O'Hara's feet for hours, holding them against his body for warmth until they began to soften. He sprinkled sulfa powder into the cracked skin to fight infection. Within a day, O'Hara's feet turned a mottled, multicolor mess, as blood and feeling returned. To relieve Monteverde, fellow crew members took hours-long shifts rubbing the navigator's feet. The process could be excruciating. O'Hara's feet didn't hurt when they were frozen; frostbite by itself isn't especially painful, because the flesh becomes numb. Burning pain accompanies the return of blood.

O'Hara's plight put a scare into every man among them. Crewmen who'd been wearing leather dress shoes traded them for rubber or leather flight boots they found in the wreckage. While inside their shelter, they aired out their boots as much as possible. Then they filled them with loose-fitting parachute silk, which provided insulation while allowing the men to move their toes to aid circulation.

Also, they learned to leave their gloves outside in the cold. Otherwise, each time the gloves thawed they absorbed more water, making it worse when they froze again. The men began to accept the reality that they were stranded in a place so cold that frozen gloves were better than soggy, half-thawed gloves.

The PN9E crew was discovering by hard experience what they otherwise might have learned less painfully had they been issued the Army Air Forces' *Arctic Survival Manual.* "Don't wear tight shoes," the manual ordered men who crashed on icy terrain. "If the shoes you have are not loose enough to allow you to wear at least two pairs of heavy socks, don't use them. Instead, improvise a pair by wrapping your feet in strips of canvas cut from your wing covers, motor covers, or any other heavy material that may be aboard your plane." The survival manual's section on shoes added ominously: "If rescue fails, your feet will be your only means of travel, so take care of them."

Unfortunately for O'Hara, without the manual the crew didn't know that rubbing his feet might worsen his condition by damaging the frozen tissue. "Don't rub the spot," it warned. "Even the gentlest massage can do a great deal of harm." The manual recommended that frostbitten feet and hands be wrapped to allow for gradual warming. The PN9E crew made matters worse by rubbing snow on each other's frostbitten skin, a useless home remedy that did nothing but make the area colder and raise blisters. Paul Spina had so much snow rubbed on him that blisters the size of tennis balls erupted on his skin.

Still, even with his broken wrist and frostbite, Spina was better off than O'Hara. The danger facing O'Hara, and to some extent all of them, was the frightful progression from frostbite to dry gangrene, a hideous condition descriptively known as mummification.

Dry gangrene is slower-acting than wet gangrene, in which bacteria infect a wound and kill surrounding flesh. If left unchecked, wet gangrene spreads through the blood with deadly results, usually within days. Dry gangrene is marginally less cruel. It results when oxygen-bearing blood can't reach part of the body. When dry gangrene takes hold, body parts shrink and turn colors. They display the reddish black of mummified skin, and then they die. The process might take years for a heavy smoker with ruined circulation. A young navigator with frozen feet might suffer the same torment within weeks.

Although unaware of the survival manual's frostbite protocols, Monteverde and Spencer did follow some of its recommendations. Many seemed to have been borrowed from a Boy Scout manual, a document familiar to both men. Without being told, the two officers kept the plane clear of snow, to improve its visibility from the air; they also worked to keep their crew well rested and well hydrated.

But other parts of the manual might have been useful had they known them. It urged downed fliers to drain the plane's oil for a smoky signal fire to be kept burning day and night. It also recommended that they remove shiny pieces of metal called cowl panels from the engines, and then place them on the wings as sun-catching reflectors to attract search planes. The PN9E crew did neither.

The crew couldn't have followed another piece of the manual's advice even if they'd known it: "DON'T GROW A BEARD if you can help it—moisture from your breath will freeze on your beard and form an ice-mask that may freeze your face." With no way to shave, all soon sported moisture-catching whiskers.

Other parts of the manual wouldn't have been useful at all. Warnings about Arctic mosquitoes applied to crashes in the sum-mer months. Blazing a trail in thick woods wouldn't be an issue, either, as there wasn't a tree for a thousand miles. Avoiding the stomach-turning qualities of parsnip root wasn't a concern; noth-ing grew on their glacier. No native Inuit people were around, so they didn't need the simple phonetic dictionary of "Eastern Eskimo" phrases such as "Where is there a white man? = Kah-bloon-ah nowk." Above all, the manual's cheerful promise that "you can beat the Arctic" by staying dry, warm, and well rested, and by eating plenty of fat, would have seemed a pitiless taunt to the half-frozen men living in a ripped-open fuselage and stretch-ing their meager rations.

Hemmed in by crevasses, the men of the PN9E instinctively followed the survival manual's most urgent command: "If you were on your flight course when you were forced down, stay with your plane. Rescue planes will be out looking for you and will find you. But remember—any search takes time. Don't give up hope of rescue too quickly. The men who are out looking for you are trained in their jobs, and if it is humanly possible to find you and get you out, they will do it."

By coincidence, on November 10, one day after the PN9E crashed, another Allied military plane went down on Greenland's east coast, a crash unrelated to the searches for McDowell's C-53 or Monteverde's B-17. Added to Greenland's scorecard for November was a Douglas A-20 attack bomber being flown by a three-man Canadian crew.

That crew would violate almost every instruction in the Arctic survival manual, with surprising results.

7

A LIGHT IN THE DARKNESS

NOVEMBER 1942

LIKE THE MEN of the PN9E, David Goodlet, Al Nash, and Arthur Weaver were a ferrying crew. Members of the Royal Canadian Air Force, the trio was supposed to deliver a forty-eight-foot, twin-engine A-20 attack bomber from Newfoundland to England, with a fuel stop in Greenland.

Flying over water two hours after they left Newfoundland, the Canadians ran into thick fog and misfortune. First, the radio went dead. That prevented radioman Arthur Weaver, dark, compact, and handsome, from checking their course or sending a distress signal. Next, navigator Al Nash, tall and lean, with a Tintin-like pompadour, found it impossible to see through the fog to plot their position by sextant. Completing their woes, pilot Dave Goodlet, aristocratic-looking, with a high forehead and a cleft chin, struggled with ice-coated wings that made it impossible to gain enough altitude to fly above the weather. Flying lower wasn't an option, either. Goodlet knew that Greenland lay ahead, and he didn't want to slam into a fog-covered mountain. They flew onward at fifteen thousand feet, fear rising as they barreled off course for hours through the soupy haze.

With a half hour of fuel remaining, the twenty-two-year-old pilot concluded that he'd run out of options. Braving whatever might be hiding in the fog, Goodlet brought the plane lower to find a place to set down. Nash provided gallows humor by narrating the descent as though they were in a department-store elevator: "Fifth floor, ladies' wear, lingerie, and fancy hosiery." They broke through the fog at thirty-eight hundred feet and saw the east coast of Greenland below—they'd crossed nearly the entire island. By then, Nash's narration had reached the bargain basement.

Goodlet estimated that they were about fifteen miles inland. He brought the plane down to five hundred feet and saw crevasses slicing across snow-covered glaciers that sloped toward the sea. Though fearful of landing nose-first in a crevasse, Goodlet knew the exhausted fuel gauge left him no choice. He slowed the plane to 110 miles per hour and kept the wheels retracted for a belly-down landing. Somehow, he threaded the needle between crevasses and landed the bomber in one piece in deep snow. When the plane shuddered to a stop, all three men were unhurt. Nash and Weaver pounded Goodlet on the back, shouting, "Good show, old cock!"

Eager to look around, Goodlet stepped outside and sank into snow up to his crotch. His crewmates pulled him inside and slammed the door. Goodlet was from Ontario, Nash from Winnipeg, and Weaver from Toronto, so they knew their way around winter. But this was something else entirely.

After sunset, the cockpit thermometer registered 34 degrees below zero and falling. The plane shook from the wind; the bomber's airspeed indicator told them the storm was blowing sixty-two miles per hour. The chopped meat sandwiches they'd brought and the coffee in their thermos bottles had frozen solid. The trio sucked at the corners of the sandwiches until they were soft enough to nibble. Their only other food was a box of hard iron biscuits, packed with nutrition but with a taste like sawdust. They had enough for eight days, if they rationed the half-inch-square biscuits one per man every twenty-four hours. Prisoners of war ate better. For warmth, they wound their parachutes around their bodies like mummy wraps. At regular intervals, they slammed their hands and feet against each other and the floor, to promote blood flow. They spent that first night in the navigator's compartment in the plane's tail, piled atop one another to share body heat. They alternated positions so each man could take a turn in the middle of the human sandwich.

Unable to sleep, Goodlet, Nash, and Weaver talked to pass the time. They discussed Gandhi, noting that the slight revolutionary had gone long periods without eating. If he could do it, they decided, so could they. Goodlet passed around a photo of his five-month-old daughter. The other two studied it so long that Weaver declared that he could pick her out from all the babies in the world. Nash confided in his partners about a girl from Michigan he'd been dating, and about his worries for his newly widowed mother. Weaver gave a blow-by-blow account of his recent wedding, lingering on the shape and cut of his bride's dress. He described his plans to build her a house, down to the last nail.

Food was a frequent subject. They told stories of Christmas dinners, and they raised their right hands and vowed never to leave anything untouched if they ever saw a full plate again. The bomber carried a cargo of cigarettes, so they smoked like fiends. Nash had been a nonsmoker, but he picked up the habit fast.

During the first two days, Weaver fiddled with the radio, with no luck. Late on the third night, the wind died down enough for Nash to take his sextant outside and calculate their position. He placed them fifteen miles inland from the Atlantic and 110 miles from any airbase or known location on their maps.

Driven by courage, self-preservation, optimistic youth, hypothermia, fear of dying without trying, or some combination, the three Canadians hatched a plan. Using Goodlet's pocketknife, they cut plywood cargo boxes and the pilot's seat into crude snowshoes. Then they inflated the plane's rubber dinghy, intending to drag it fifteen miles over snow and around crevasses to reach the water. Once there, they intended to climb in and paddle one hundred–plus miles to the nearest settlement. As implausible as the plan sounded, they concluded that it was their best chance. They gathered a flare gun with a box of cartridges, three pyrotechnic marine distress signals, their iron biscuits, and all the cigarettes they could carry.

Before setting off, Weaver tried the radio one last time. Unexpectedly, a weak signal reached a Canadian airport. Weaver sent three SOS calls and gave the position Nash had calculated by sextant. His fingers froze, so he pounded his coded message on the transmitter key with his fist. On the third SOS try, before the batteries gave out, the airport responded that the message had gone through.

Expecting that help was on the way, they postponed their journey. But after two days, with no sign of rescuers and their biscuit rations reduced to one-quarter a day, they stopped waiting. They inflated the dinghy, destroyed their plane's bombsight, and burned all papers that might be useful if an enemy happened upon the abandoned bomber.

Goodlet, Nash, and Weaver leaped into the wind and snow but didn't get far. After two hours of pulling the dinghy, they'd traveled a quarter mile. Dejected, they turned back and spent the night in the plane, smoking cigarettes and smacking each other's arms and legs to keep blood circulating. Their mouths grew bloody and sore from sucking shards of ice and snow.

The next day, Greenland's weather took a strange turn: the temperature soared by about 50 degrees Fahrenheit, heated by a warm breeze called a *foehn* or "rain shadow" wind. The warmth made for mushy trekking, but it encouraged Goodlet, Nash, and Weaver to try again, this time with no turning back. They made good progress, but when night fell, Greenland's punishing cold winds returned, freezing their flying suits to their bodies like icy armor. For the next seventeen hours of darkness, they huddled under the inflated dinghy, praying. Weaver promised to become a regular Sunday churchgoer if he ever again found himself in a place with churches. They'd slept just a few hours since the crash, yet none of them could nod off.

The following morning the weather broke and the skies were clear. They resumed their journey at daylight, heading east toward the coast. While veering a mile off course to avoid a crevasse, they heard the unmistakable sound of an airplane engine. The Canadians dove into the dinghy for their supplies. The flare gun proved useless—the firing mechanism had broken in the cold. The first marine signal was a dud. So was the second. But the third and last one worked, lighting the sky. The search plane wagged its wings in salute. The date was November 18, 1942, eight days after the crash. They'd been found.

The search pilot reported that the men had traveled a remarkable seventeen miles northeast on foot from their downed plane. He circled low over them. Soon the trio saw small parachutes open, drifting to earth like milkweed seeds. The crates beneath the chutes carried food, clothing, sleeping bags, snowshoes, one hundred feet of rope, and a bottle of Scotch.

Nash had never touched liquor, but just as he'd become a smoker, now he grabbed the Scotch. Half mad with thirst, he plucked out the cork and drank eight ounces—on an empty stomach and no sleep for days. Within minutes, Nash had sprawled onto his bottom. His eyes rolled around his head—Weaver thought they looked like marbles in a milk bottle. Down for the count, Nash slumped onto his side and passed out. His companions couldn't wake him, so they dressed him in dry clothes and a parka, then stuffed him in a sleeping bag. Goodlet and Weaver put on dry clothes, too, then gorged on K rations from breakfast to dinner. After an hour of sleep, they woke up retching, having overwhelmed their shrunken stomachs. Nash woke and followed them down the binge-purge path.

The Canadians found a note among the supplies instructing them to tie themselves together and continue toward the water. The men's northeasterly path was taking them toward a notch in Greenland's coast known as the Anoretok Fjord.

Based on the search pilot's report, a plan emerged for the U.S. Coast Guard cutter *Northland* to fight through the ice and meet the Canadians at the fjord. Before joining the rescue effort, the *Northland* had been transporting freight and bringing about eighty U.S. soldiers to the new airbase at Bluie East Two.

Why the *Northland* hadn't been involved in the rescue effort sooner—not only for the Canadians, but also for McDowell's and Monteverde's lost American crews—was never explained. One possible reason was the ossified competition among military branches. Officials of the U.S. Army Air Forces were overseeing the C-53 and B-17 searches, and the service had recently established rescue stations along Greenland's east coast. A successful rescue would prove the value of the stations and the skills of army men, with bragging rights as a bonus. The army had no incentive to hand the Coast Guard the mission, and the potential rewards, unless absolutely necessary.

As the *Northland* headed toward a rendezvous point with the Canadians, the pilot of the ship's amphibious Grumman Duck took flight. Hoping to make sure the three frozen travelers were headed toward the correct fjord, Lieutenant John Pritchard scoured the ice cap. Pritchard spotted the men's trail of snowshoe tracks, which he thought looked less than two days old. But despite one pass after another, he couldn't find Goodlet, Nash, and Weaver.

Bound together, the Canadians used the airdropped snowshoes to quicken their pace. They stopped when it got dark, afraid of yawning crevasses, only to be drenched by sleet and rain. Rather than wallow in slush, they stood through the long night like napping horses, holding sleeping bags over their heads for protection. Heavy fog the next morning held the same risk of crevasses, so they spent much of the day massaging their feet, using the Scotch as rubbing alcohol. When the sky cleared, they continued eastward toward the sea. Too exhausted to stand, that night they lay down on the ice with their arms wrapped around each other. Soon they were frozen together, and it took all their fading strength to pry themselves apart. Once separated, Weaver pulled off his right boot and found his foot had frozen solid, leaving the skin white, waxy, and numb.

The closer the trio got to the water, the more crevasses they encountered. They felt the glacier heaving beneath their feet and heard the thunder of icebergs calving nearby. Doubts rose in their minds whether they'd survive. Nash suggested that they sing a hymn but they didn't know one. Instead they crooned "God Save the King" and a hit song, "Praise the Lord and Pass the Ammunition," about a chaplain who took up arms during the Pearl Harbor attack. Their lips bled as they sang, but it raised their spirits through the night.

The next morning, a hard crust on the glacier made their snowshoes feel like skates as they glided toward the water. Looking out to the sea, they spotted what looked like a rowboat. It was the *Northland*, perhaps ten miles from shore, too far for the Canadians to attract the crew's attention.

Setting aside their pain and hunger, their thirst and exhaustion, they raced the last two miles to a sheer ice cliff at the water's edge. Using a lighter he'd filled with alcohol from the plane's radiator, Goodlet tried to set fire to their parkas to signal the ship. But the coats were too wet to burn. The three men were literally and figuratively at the edge of a cliff. Their flare gun was broken, their marine signals spent. They had nowhere to go. Their one hope was that the crew of the *Northland* would see them. The sky was clear and moonlit, filled with stars beyond number.

The *Northland* fired flares and illuminating shells, a light show that reminded Weaver of fireworks for the queen's birthday. The ship came closer to shore and swept its powerful spotlights back and forth along the coast. The trio danced and waved their parkas each time the spotlights hit them, but the lights never stopped. The men were too small and the ship was too far. The Canadians saw the Duck take flight from the water near the *Northland*, but still the trio saw no indication that their would-be rescuers had spotted them. Goodlet, Nash, and Weaver watched as the Duck returned to the ship.

THE *NORTHLAND* IN ICE DURING WORLD WAR II. *(U.S. COAST GUARD PHOTOGRAPH.)*

The *Northland* was in dangerous waters, in heavy pack ice with icebergs and growlers all around. With no sign of the lost men, the *Northland*'s captain, Lieutenant Commander Frank Pollard, had to focus on the safety of his own men. The Canadians stared into the darkness as the ship steamed away from the coast.

Thirteen days had passed since their crash, five since they'd been spotted from the air. They were weak, wet, freezing, and out of food. They knew they'd never survive a trek back to the plane, and they saw no point in trying. They doubted they'd live through even one more night of 40-below-zero cold. Desperate, with nothing to lose, Goodlet thought they should try again to light their parkas. If it didn't work and the coats were ruined, death by hypothermia would come mercifully sooner.

They tore the coats into strips to help them burn. Goodlet's lighter was low on fuel, so it took repeated spins of the wheel to raise sparks. When it lit, he touched the flame to the parka strips. This time the fabric caught fire, and the men stoked the remnants of their coats into a glowing, smoky blaze. But soon the flames died into darkness.

The *Northland*'s crew hadn't wanted to give up on the Canadians. Some kept looking back toward shore, hoping for a sign of life. Before the little parka fire went out, it caught the eye of the ship's chief gunner's mate, who'd been watching through binoculars.

"I just saw a light," he told Ensign Charles Dorian, who rushed to the bridge to tell the captain. Pollard spun the *Northland* back toward shore. He ordered his men to turn on the ship's big searchlight and shoot a half-dozen "star shells" that turned night into day as they fell to earth.

At the edge of the ice cliff, the three men yelled with joy and relief. Giddy and renewed, they pounded on each other's backs. Weaver read a Morse code message from the ship's blinking signal lamp: "Move back from edge of glacier and bear south to meet landing party."

On the *Northland*, Pollard faced two decisions: who would carry out the rescue and how it would be accomplished. He settled on a plan that called for a small team to pilot a motorboat through the ice-filled waters, reach the shoreline, climb a glacier, avoid crevasses, guide or carry the three men back down the glacier, and get them to the ship in one piece. The dangers were too many to count, but no better ideas emerged.

Frustrated that he hadn't been able to find the Canadians in his Duck, John Pritchard volunteered to lead the mission. Equipped with skis and snowshoes, the lieutenant and a ten-man team reached the shore by boat, roped themselves together, and found a back way to scale the icy cliff. Pritchard led his men across a heavily crevassed section of the unstable glacier. By shouting and flashing searchlights, they located Goodlet, Nash, and Weaver. A photograph of the meeting commemorated the happy occasion.

With darkness upon them and the glacier pouring chunks of itself into the water, the rescuers and the Canadians rushed back in the direction of the motorboat. Pritchard led the group down the face of the cliff, which seemed intent on tossing them into the sea. When everyone was safely aboard, Pritchard and his crew steered the motorboat through the dark to the *Northland*'s side.

When the Canadians stepped aboard the cutter, they were feted, fed, and coddled so thoroughly that Weaver said they felt like newborns. Pollard, the *Northland*'s captain, told the men he'd written them off as dead before the lookout spotted their burning parkas. The ship's doctor treated their frostbite and windburns. He diagnosed the mental effects of hypothermia, a confusion that Weaver described as "twilight between sanity and insanity." The doctor told them they were within a day of cracking altogether.

JOHN PRITCHARD, FAR RIGHT WITH ROPE AND POLE, AFTER LEADING THE RESCUE OF THREE CANADIAN AIRMEN WHOSE PLANE WENT DOWN ON THE ICE CAP. CANADIAN PILOT DAVID GOODLET IS FRONT LEFT, IN A FLIGHT SUIT; TO HIS LEFT, IN A BORROWED COAT, IS FLIGHT SERGEANT ARTHUR WEAVER. NAVIGATOR AL NASH IS IN THE ROW BEHIND THEM, IN A PARKA. IN THE BACK ROW, SECOND FROM LEFT WITH A CIGARETTE, IS ENSIGN RICHARD FULLER, WHO WOULD SPEND FIVE MONTHS LEADING ANOTHER RESCUE TEAM ON THE ICE CAP. *(U.S. COAST GUARD PHOTOGRAPH.)*

On the bright side, Weaver said, their blurred judgment had allowed them to persevere when logic, hunger, thirst, and exhaustion might otherwise have made them curl up on the ice and die.

All three were thin and hollow-eyed. Sleep would be a problem for the foreseeable future; they'd startle awake from shivering nightmares in which they were back on the glacier. But they were safe.

The Canadian trio spent the next six weeks aboard the *Northland*, celebrating their unlikely survival and regaining their health. Later they told their story to reporters, posed for photos, and saw their tale recounted in magazine stories and a comic book called *Lost in the Arctic*.

Asked what kept them going, Weaver said, "Dave had his wife

and baby daughter. Al was worried about his mother, alone out in Winnipeg. And I had my wife. Do you see what I mean? We had something to live for."

John Pritchard's heroism didn't go unnoticed. The unassuming young lieutenant was a Coast Guard search pilot, yet he captained a motorboat and climbed a glacier to rescue three men, jeopardizing his own life. He and the *Northland*'s captain, Frank Pollard, each earned the Navy and Marine Corps Medal, the second-highest noncombat award for bravery.

Pritchard's citation read, in part, "Lieutenant Pritchard's intelligent planning, fearless leadership, and great personal valor aided materially in the gallant rescue of the stranded men, and were in keeping with the highest traditions of the United States Naval Service."

IN THE TAIL section of the PN9E, no such celebrations were under way.

The men of Monteverde's crew didn't know that the Canadians had walked off the ice, or even that they'd crashed. It wouldn't have mattered. After Harry Spencer's fall into the crevasse, walking toward the water wasn't an option for them.

On the other hand, Monteverde, Spencer, and O'Hara discussed whether some or all of them might walk to a weather station in the opposite direction from the water that they'd noticed on a map salvaged from the cockpit. Spina joined the conversation and pronounced the idea suicidal. He told them he'd rather remain with the plane than freeze to death along the way. After a long talk, everyone agreed.

For military planners, the *Northland*'s success in reaching Goodlet, Nash, and Weaver demonstrated that the Coast Guard needed to be more heavily involved in the search for Monteverde's B-17 and McDowell's C-53, lost for fourteen and eighteen days, respectively. After the Canadians' rescue, the ship received a congratulatory message from "Iceberg" Smith, the rear admiral who commanded the Greenland Patrol: "Well done. Suggest *Northland* proceed . . . for search of Baker Seventeen [the B-17 PN9E] and . . . Cast Five Three [the C-53]."

The two American crews still on the ice cap hadn't been forgotten, and Pritchard and the Duck might have another chance to bring lost men home.

8

THE HOLY GRAIL

JANUARY 2012

IF THE PENTAGON is a battleship, Coast Guard Headquarters is a tugboat. Located near the mouth of the Anacostia River, in a far southwest corner of Washington, D.C., the headquarters building is like the service itself: modest, practical, without flash or self-importance.

The main lobby resembles the entrance to a struggling small-city hospital. On this bright winter day, the smell of glue is over-powering, wafting from a nearby hallway where workmen repair broken floor tiles. A portly janitor swabs the floor. A bored security guard leans back in his chair and talks movies with a colleague.

Into the lobby blows the Duck Hunter, Lou Sapienza. He's here to press his case yet again with Defense Department officials for money to support his plan to lead a team to Greenland, find the lost Duck, and recover the remains of its three occupants. Lou's unmistakable voice precedes him, bouncing off the walls as he saunters through the lone metal detector.

More so than on his last visit to Washington, three months

ago, Lou is in seemingly friendly waters here. The Duck's heroic story is braided into Coast Guard lore, and the service's top brass would like Lou to succeed. The headquarters building, tired as it may be, is home to an elaborate scale model of the little plane, and talk has percolated here for years about exhuming the real thing from under the ice cap.

Still, money is tight, and there's only so much the Coast Guard can do. That's why Lou's real audience today are military and civilian members of the Department of Defense Prisoner of War/Missing Personnel Office, DPMO, the same men and women who poured cold water on him in October.

As a MILITARY service, the Coast Guard is something of an odd duck itself. It has a Swiss Army–knife mission of law enforcement, humanitarian relief, and military tasks, including coastal security, drug interdiction, search and rescue, marine safety, and environmental protection. Those roles aren't a natural fit for the Defense Department, so the Coast Guard has bounced over the years from the Treasury Department to the Transportation Department to its current home in the Department of Homeland Security. And yet, at times of war, the Coast Guard can be swallowed by the navy, as it was during World War II.

Unfortunately for Lou, pretty much the one job not included in the Coast Guard's mission statement and $10 billion annual budget is recovering lost World War II airmen and missing biplanes. That work is funded by a tiny fraction of a sliver of the U.S. Defense Department's $525 billion annual budget.

Still, Coast Guard officials have offered to consider using a C-130 Hercules transport plane to haul Lou, his team, and the necessary equipment to Greenland. The service also has provided advice, research assistance, general support, and a boardroom for this morning's meeting.

As he readies himself for his pitch, Lou seems oblivious to the skepticism that his contacts at DPMO have expressed in e-mails about the million-dollar price tag for a mission to find the Duck as well as McDowell's C-53. Lou seems equally unaware that DPMO's entire projected budget this year is about $22 million, and its staff consists of just forty-six military personnel and eighty-seven civilians. Maybe worst of all, DPMO's already approved budget calls for the office to "deploy investigation teams to Serbia, Poland, Czech Republic, Romania, and Tunisia." Nowhere within a thousand miles of Greenland.

What Lou doesn't know can hurt him, but at the moment he's in high spirits. His zip-up fleece jacket is adorned with five brightly colored patches, the most prominent one for the Duck Hunt. Another displays the logo of his expedition company, North South Polar Inc. Two more trumpet other potential recovery missions, one in Greenland for McDowell's C-53, and one in Antarctica for a lost World War II–era navy plane called the *George 1*. The fifth patch is for the Fallen American Veterans Foundation, a nonprofit organization Lou created to pursue corporate and private sponsors who share his vision of recovering American MIAs from around the world.

"Hey there!" he calls. I'm pressed against the patches in a hug.

Since the last Washington meeting, Lou has been a round-the-clock dervish of research and logistics. On the research front, his priority has been to pinpoint where to dig and melt through the ice for the Duck and its men. To narrow the search, he's been comparing historical documents and clues against data from modern technology, including sensor findings from NASA survey flights and ground-penetrating radar from U.S. military planes on their way home from Afghanistan and Iraq. He's become an amateur authority on the flow of glaciers, knowing that in Greenland, anything from a pebble to a plane moves along with the ice in which it rests. Lou has also learned how ice and snow build up over the years in different parts of Greenland, which leads him to estimate that after seven decades at Koge Bay, the Duck is thirty to fifty feet below the surface.

On the logistical side, Lou's been pricing helicopter time from Air Greenland, evaluating cold-weather gear from possible sponsors, and choosing freeze-dried foods from a company beloved by apocalypse-minded survivalists. Its nitrogen-packed, enamel-coated cans of beef stew promise to taste good until 2037. I worry that it might take that long to raise the money needed for the Duck Hunt.

Although Lou is point man for the search effort, he's not its natural father. He adopted that role from a retired Coast Guard captain named Tom King, a barrel-chested, sixty-year-old fireplug of a man who suffered from a recurring nightmare about eBay.

KING GREW UP in Harrisburg, Pennsylvania, with boyhood hopes of attending the U.S. Naval Academy at Annapolis and captaining a submarine. By the time he assembled the required recommendations, the seas had shifted and he was instead headed toward the Air Force Academy. But his heart wasn't in it. His high school German teacher suggested the Coast Guard. At seventeen, King caught sight of the *Eagle*, the Coast Guard's three-masted cutter, known as "America's Tall Ship." A service that sailed such a magnificent ship was the place for him.

During a thirty-year career, King rose to chief of the Coast Guard Office of Aviation Forces. After retiring in 2004, he launched an aviation and homeland security consulting firm. He also became involved with the Coast Guard Aviation Association, a fraternal group known jovially as the Ancient Order of the Pterodactyl, a title befitting members' proud self-image as flying dinosaurs. Their motto: "Flying Since the World Was Flat."

His work with the Pterodactyls gave King time to reflect on the case of Lieutenant Jack Rittichier, the first Coast Guardsman killed in Vietnam. In June 1968, piloting a combat rescue heli-

copter known as a Jolly Green Giant, Rittichier was hovering over
an injured Marine Corps pilot when North Vietnamese troops
opened fire. He tried to land, but his craft exploded on impact.
At the war's end, Rittichier was the only member of the Coast
Guard still declared missing in action from that conflict. A quarter century passed before a joint American-Vietnamese search
team found his remains. In 2003 Rittichier was buried with honors on Coast Guard Hill at Arlington National Cemetery, a hal-
lowed section normally reserved for top commanders. Delivering
the eulogy, Coast Guard commandant Admiral Thomas Collins
declared, "All hands are now accounted for."

That was true for Vietnam. But it troubled King that the same
couldn't be said for Coast Guardsmen from World War II, notably
the pilot and radioman of the *Northland*'s Duck, John Pritchard
and Ben Bottoms.

King's focus on the Duck intensified in 2007, when he learned
that a Coast Guard Academy ring was being auctioned on eBay.
The ring had belonged to the late Captain Frank Erickson, a
Coast Guard giant who pioneered the use of helicopters on rescue
missions. King shuddered at the thought of Erickson's ring being melted down or worn as a golden bauble on some rich guy's
knuckle. King and his friend Captain Mont Smith, president of
the Pterodactyls, won the eBay auction for $2,025. They carried
the ring to Coast Guard events and reunions, treating it like Cin-
derella's shoe by allowing fellow aviators to try it on for size.

Coast Guard relics became King's passion, and he was named
the Pterodactyls' vice president for museums, aircraft, artifacts,
and restorations. By the time he and Smith won Erickson's class
ring, King had sent an e-mail to fellow Pterodactyls asking what
they considered to be the Holy Grail of Coast Guard aviation artifacts. King knew the answer before the survey was done: John
Pritchard's Grumman J2F-4 Duck, serial number V1640, lost in
Greenland.

Certain that the Duck should be displayed in the Smithsonian or a Coast Guard museum, King felt sickened by the possibility that a wreck hunter might recover the plane from the ice and sell it to a private collector of warbirds, as vintage military aircraft are known. King knew that the lucrative market for warbirds had led salvagers to seek them out in jungles, on mountaintops, beneath the seas, and on glaciers. Only thirty-two Grumman J2F-4s had been built, and only one of those Ducks remained in flying condition, making it one of the rarest warbirds. Because Pritchard's Duck carried a heroic backstory, it could be worth several million dollars on the private market. Even more horrifying to King was the possibility that a wreck hunter might disturb the human remains and grave-rob personal items.

"I don't want to see John Pritchard's wallet being sold on eBay," King says, shaking both his head and his fist. "We can't allow that."

The Coast Guard did try to reach the Duck in 1975, using two helicopters and a shore party. Conditions for the search were ideal, but they were looking in the wrong place. As King talked up the idea of a new search, he got word that a private collector had already obtained permits from Greenland to launch an expedition to retrieve Pritchard's plane. King's eBay nightmare seemed to be coming true.

Back-channel messages were sent to Greenland government officials that the remains of American World War II casualties were believed to be inside the plane. That made the Duck an overseas American military gravesite. Wreck hunting was one thing; disturbing the resting place of heroes was something else entirely. The wreck hunter's permits were squashed. Still, it was a close call. Considering the Duck's potential value, other wreck hunters might start circling, perhaps without going through the formal permitting process. Fears about the Duck's fate rose further amid reports of rapid melting of the Greenland ice cap. The plane and its occupants might be exposed to the elements and treasure seek-ers both. A sense of urgency took hold.

King enlisted his friend Mont Smith, the Pterodactyl president, and Smith buttonholed Vice Admiral Vivien Crea, at the time the Coast Guard's second-ranking officer. Crea held an exalted posi-tion as the Pterodactyls' "Ancient Albatross," an honorary title bestowed on the service's longest-serving aviator. Crea passed the mission of saving the Duck to Captain Mike Emerson, then the Coast Guard's chief of aviation.

In February 2008, Emerson walked into the aviation office on the third floor of Coast Guard Headquarters. Pausing at a cluster of cubicles, he dropped a pile of papers on the desk of one of his project managers, Master Chief Petty Officer John Long.

"Hey," Emerson said, "see if you can find something out about this." The papers contained the outlines of the Duck story.

Long and Commander Joe Deer, later replaced by Commander Jim Blow, spent the next two years working as historical sleuths, haunting archives to dredge up declassified reports, details, maps, photographs, obscure references, and even rumors about the plane, its men, and their final flight. They pored over radar data and stud-ied the movement of glaciers. They tracked down family members of the lost Duck crew. One goal of those contacts was to discover how family members felt about a mission to bring the men home. Another was to collect DNA samples from relatives of the lost men, to prove their identities if bodies or bones were found.

In 2008 and 2009, radar and sonar scouting missions identified what seemed like a promising place to dig, and an expedition took shape for 2010. When Coast Guard officials began looking for help from outside experts to carry out the mission, they teamed up with a guy named Lou.

THE ELDEST OF four boys, Luciano "Lou" Sapienza spent his childhood in suburban New Jersey, thirty miles from Manhattan.

His father, a World War II navy veteran, worked as an import-export manager for a brewery supplier. His mother founded the Somerset County Association for Retarded Citizens and served as its executive director, a role that fit her work helping one of Lou's brothers, who was developmentally disabled.

As a child, Lou was quiet and withdrawn, "a little bit of a mama's boy." Beyond his driveway were woods where he'd have imaginary adventures until he heard his mother's booming voice, "scaring the hell out of me" and calling him home. He went to parochial school, where he won a camera as a prize in a third-grade magazine drive. "I was tied with this girl, and Sister Mary Lawrence told me, 'Be a gentleman and let her have the camera.' I said 'No.' I really wanted that camera." The nun thought of a number between one and ten; Lou picked three and won. He walked away with a box-shaped Imperial Satellite 127 camera and his first calling.

"I'm one of those people who always feel as though I was born in the wrong generation," Lou says. "Photography for me, in retrospect, has always been about preserving the past. I have this thing about the past."

Before embracing photography, though, he had been on a crooked path toward the priesthood. He attended a seminary during freshman and sophomore years of high school, but the priests gave him the option of leaving on his own or being thrown out. "It had something to do with girls. Nothing major," he says. Lou ended up at an all-boys parochial high school where he had few friends and felt most comfortable behind a camera. "Photography helped me. It helped me to relate."

After floundering through two years as a communications and theater arts major at a small college in Indiana, Lou returned home. He took photography courses at Cornell Capa's International Center for Photography in Manhattan, at New York University, and at the Maine Photographic Workshops. A career as a commercial photographer followed, along with marriage and three sons.

"In 1989, I was married and we'd just bought a house in Plainfield, New Jersey. I always had CNN on in the background. I was walking from one room to another, and I heard that a group of American explorers were going to Greenland to search for the Lost Squadron. I was stuck and bogged down doing commercial stuff. This was the type of story I always wanted to tell. Also, I always wanted to do something adventurous. Not just adventure for adventure's sake. I was fascinated with World War II. This was about rediscovering the past, adventure with a purpose. I called CNN and tried to find out who these people were."

He tracked down Norman Vaughan, a renowned explorer who'd been a dogsled leader during Admiral Richard Byrd's 1928 expedition to Antarctica. During World War II Vaughan served as a lieutenant colonel, and in July 1942 he took part in the rescue of the twenty-five men of the Lost Squadron. When Lou began

LOU SAPIENZA IN GREENLAND IN 1992 WITH A .50-CALIBER MACHINE GUN FROM *GLACIER GIRL. (COURTESY OF LOU SAPIENZA.)*

searching for Vaughan, the explorer was past his eightieth birthday yet still competed in Alaska's Iditarod dogsled race. In his
spare time, Vaughan had joined a team that intended to retrieve
one of the Lost Squadron's six P-38 Lightning fighter planes, buried under more than two hundred feet of snow and ice.

Lou wrote a passionate letter making his pitch: he would serve
as the expedition's official photographer and do whatever else was
needed, including cooking and hauling equipment, to take part
in the mission. It worked. "They told me, 'We've got a lot of good
photographers who want to go. You wrote a better letter.'"

"My first trip to Greenland," Lou says, "we were out there about
fourteen days. I was told, 'Wear what you normally wear when
you go skiing.' I had never been skiing. I wore jeans and imitation
Sorel boots, and I started going hypothermic. The boots soaked
through. I had to wrap my feet in plastic bags. I learned that if
your feet ain't happy, you ain't happy. When I was invited to go
back to Greenland the next time, I got sponsors, and we had two
pairs of real boots for everybody."

When Lou joined the team, Vaughan and other expedition
members had already made four unsuccessful trips to Greenland
in pursuit of their prize. Lou participated in the last three expeditions, including a climactic 1992 effort in which they drilled and
melted through almost a football field of ice, to a depth of some
268 feet. They created an otherworldly ice cavern around a P-38
Lightning, disassembled it, and hauled it piece by piece to the surface. That plane became *Glacier Girl*, and Lou's photos became the
visual record of the expedition.

Lou returned to his life, got divorced, had a long relationship,
broke up, met someone else, got married and divorced again.
Through it all, he never shook the idea that he had found his true
calling. He read a magazine article about three navy fliers buried
in 1946 under the ice in Antarctica, inside a flying boat called the
George 1. Lou located the men's families and offered his services. "I

told them, 'We've done this before. I know how to get them.'" The navy still hasn't signed on, saying the search would be too dangerous. But Lou won't give up, which is why the *George 1* mission is embroidered on one of his patches.

"If you tell me it can't be done and I know it can be done, I don't take that kind of answer very lightly," Lou says. "I feel like I can't turn my back on the families. As long as I'm alive, these men are coming home."

While promoting his plan for a *George 1* expedition, Lou presented his case to the government's Joint POW/MIA Accounting Command, or JPAC, which works with DPMO to recover unaccounted-for American service members. JPAC wasn't ready to pay for the *George 1* project, but through a roundabout series of contacts, the agency put him in touch with the Coast Guard's Duck Hunt team. A public-private partnership was born.

"We went from there," Lou says. "They wanted me to provide positive, verifiable proof that we knew where the Duck was." They

GLACIER GIRL SOME 268 FEET UNDER THE ICE CAP. *(COURTESY OF LOU SAPIENZA.)*

went to Greenland in September 2010, where Lou and his team investigated a site that a previous Coast Guard contractor had identified as the likely resting place for the Duck. That location proved to be a dud, so Lou regrouped.

He returned home to a town on eastern Long Island, New York, where he lives with his twenty-two-year-old son, Ryan, and his dachshund, Sarge. Convinced that he could succeed, Lou deepened his research and focused on winning support to finish the job. As months passed with no sign he'd get a green light, Lou forced himself to remain upbeat. In private moments, though, he'd acknowledge that several years had passed since he'd earned a steady paycheck. The little money he had left was going into planning the Duck Hunt. He was surviving on savings, an inheritance, a supportive girlfriend, and faith.

"The biggest thing is faith," Lou says. "Faith in the national commitment to leave no man behind.

"One of the things I've realized is that this screwy life that I've had was all leading up to being able to put an expedition together to Greenland," he says. "I bring certain skill sets to this that a lot of people don't have, and I'm able to get it done. If they're there, we will find them. If the families and the Coast Guard want these men home, that's good enough for me. It's the ultimate way to honor these men and what they did to bring them home to their families."

9

SHORT SNORTERS

NOVEMBER 1942

Stormy weather returned to Koge Bay from November 13 to 16, the four days following Harry Spencer's rescue from the crevasse. Driving sleet and snow kept the nine men of the PN9E trapped inside the remains of their bomber. Wedged together in their cocoon of silk parachute cloth, they had all day and night to think, and to worry.

They salvaged what they could from the front end of the bomber, ripping up the floor in the radio compartment and moving it to the tail to create more room for sleep. They tore out cabin insulation for bedding. They dug through the snow near the wreck and found crew members' personal belongings, including several garment bags containing clothes, cigarettes, candy, and gum to share.

With time to kill, they invented a primitive calendar. To mark the month, they lined up a row of eleven matchsticks in a pile of snow inside the bomber tail. Below that row, they lined up nine matchsticks, to note the day they crashed. They added one matchstick for every day after that. Several of the men had heard that no one had ever survived on the ice cap for longer than two weeks, so they expected to be rescued or dead before they added a twelfth matchstick to the top row.

They ventured outside in the mornings to clear off heavy loads of snow that piled up on the wings overnight. They didn't know it, but there was little point to the exercise: the weather made it nearly impossible for searchers to take flight. One search flight went up on November 13; no planes flew the next day; two were airborne on the fifteenth; and none searched on the sixteenth.

The foul weather also meant no search flights for McDowell's missing C-53.

Daily logs from Bluie West One recorded the frustration: "November 13, 1942: We are unable to continue search today on account of weather. . . . November 14, 1942: Search for lost plane could not be continued today on account of weather. . . . November 15, 1942: We are unable to search for lost plane today on account of weather conditions and daylight shortage when weather was clear over the Ice Cap."

With no new radio communications from McDowell's crew, doubts rose among military officials whether the five men in the downed cargo plane were alive. Yet officially the search continued.

As winter approached, the nights grew longer, leaving precious little warming sunlight on the ice cap. At the December solstice, sunrise and sunset would be separated by little more than three hours at Koge Bay.

Inside the PN9E, the combined leadership of Monteverde and Spencer held things together. The men continued to pass time beating their arms and massaging their legs in futile attempts to get warm, or at least warmer. They moved around as much as possible, which wasn't much in the broken tail. They played spelling games. They recounted their life stories. Woody Puryear came to believe that he knew the eight men crowded against him better than anyone else on earth, despite having only met them a week earlier when he came aboard as a volunteer searcher.

At what passed for mealtime, they used body heat to thaw their partial rations under their armpits. They didn't have a Bible, but they spoke of God, and they continued their prayer sessions. The men came from several denominations, but all were Christian. Monteverde prayed the rosary daily, a spiritual response to what he considered banishment to a frozen white hell.

During silences between stories and prayers, Woody Puryear thought about his mother; his sisters, Blanche and Pearl; and his girlfriend, Erma Ray Yates, back in Kentucky. Like hungry survivors everywhere, he and the others talked and fantasized about food. For Puryear, the imagined meals featured old standbys such as steaks and chops, but also little sandwiches his mother made with a dab of peanut butter between two crisp crackers. Thoughts of home loomed large for Puryear, who in ten months had gone from country boy to Greenland-based soldier to missing man. The

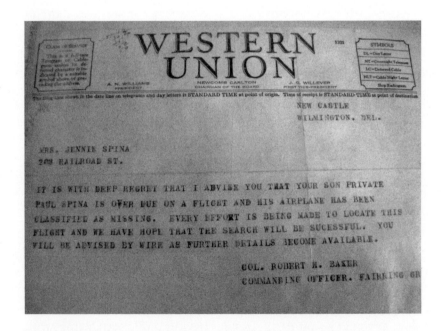

WESTERN UNION TELEGRAM RECEIVED BY PAUL SPINA'S MOTHER, JENNIE, REPORTING HIM MISSING. (COURTESY OF JEAN SPINA.)

young staff sergeant could find some relief in the knowledge that he'd already sent his Christmas cards, two months early, to be sure they'd arrive home on time.

One day, when several men seemed to be losing hope, Monteverde asked, "How many of you are Short Snorters?" Weak smiles knowingly crossed lips.

SHORT SNORTERS WERE a loosely bound society of airmen and -women who enjoyed a drink and a good time. The club traced its origins to 1925 and a handful of rough-and-ready Alaskan bush pilots with too much time on their hands. Beyond that, the group's creation stories were like bar tales: confused and often contradictory. The main requirement for induction in the Short Snorters was completion of a flight over at least one ocean, either as pilot or passenger. By virtue of that alone, a prospective member was considered a man or woman of the world, with stories to tell and friends to share them with. A "short snort" was slang for a small or weak drink, so it was said that Short Snorters were fliers who knew how to kick back but also knew the importance of keeping a clear head.

Beyond a transoceanic flight, requirements for initiation were four dollars and three relatively sober sponsors. Each new member had to be nominated by three existing Short Snorters in exchange for payment of a dollar, or a drink, to each sponsor. The fourth dollar bill became the new Short Snorter's membership card. Each sponsor signed the bill and noted the place of initiation. The new member, in turn, signed his or her sponsors' membership bills.

From then on, a Short Snorter was required to produce his or her membership dollar in reply to the inquiry, "Are you a Short Snorter?" Failure to produce the signed bill within two minutes would cost the member a dollar or a drink to all fellow members present. One forgetful Short Snorter tattooed a dollar bill on his chest because he said it was less expensive and less embarrassing to open his shirt than to pull out his empty wallet. When membership dollars were produced, Short Snorters would sign each other's bills, noting where they'd met. When a bill could hold no more signatures, its owner would tape a new bill onto its end, and then another; some Short Snorters had membership bills stretching several yards long. After a Short Snorter's death, the membership bills became a paper memorial that commemorated the people and places he or she had encountered along the way.

World War II created exponential growth for the Short Snorter movement. A *New York Times* story called it "a billion dollar racket with around three million members." Egalitarian and inclusive by design, the organization admitted women, accepted foreign currency, and inducted members from Britain, Canada, Australia, and Russia. Some German pilots got wind of the idea and created their own branch. By the end of the war, the Short Snorter web reached up to generals, among them Dwight D. Eisenhower and George S. Patton Jr.; diplomats including W. Averell Harriman; and world leaders such as President Franklin D. Roosevelt, Prime Minister Winston Churchill, King Peter of Yugoslavia, and Prince Bertil of Sweden. First Lady Eleanor Roosevelt joined, too.

If every Short Snorter bill had been collected and collated, the result would have been the membership list of a vast global social network, from peons to potentates, connected by fate, friendship, and thirst.

By asking whether his fellow castaways were Short Snorters, Monteverde had found a way to bind the lost crew of the PN9E through their ordeal and beyond. Signing each other's bills was an act of defiance. It said that survival was possible, and someday they'd cross paths and share stories of their incarceration on ice.

The crewmen dug into their wallets. Puryear couldn't find his, so he borrowed the four-dollar admission fee. They passed around the bills and signed their names. All they were missing were shots of whiskey.

AFTER DAYS OF sending distress calls on the Gibson Girl transmitter, radioman Lolly Howarth had nothing to lose except his fingers: he volunteered to go out into the cold to work on the PN9E's larger, more powerful radio equipment.

Howarth knew that it was a long shot. When the bomber broke in half at the radio compartment, most of the gear had been thrown around the cabin or left dangling from its mountings. The radio boxes, ranging in size from breadboxes to small suitcases, were now piled atop one another as a windbreak. But on closer examination, Howarth discovered that the glass vacuum tubes in several pieces had somehow survived the crash.

During the ten months since he'd enlisted, Howarth had trained hard as a radio operator, but he was no veteran expert in communications or electrical engineering—just a corporal who wanted to be an actor or a drama teacher when the war ended. Five-foot-seven and 130 pounds, the brown-eyed, baby-faced Howarth looked younger than his twenty-three years. Born in a log cabin built by his logger father, Howarth was the second of four brothers who hunted deer and lived off the land in rural northeast Wisconsin. Quiet and sweet natured, Howarth was perhaps the PN9E crewman most comfortable with his surroundings; heavy snows cut off his family's cabin for much of each winter. Howarth and his brothers had to ski several miles daily between home and school. After leaving home at eighteen, Howarth worked his way through La Crosse State Teachers College by washing dishes in local restaurants, becoming the first member of his family with a college degree. Single when he'd enlisted, now he had a wife, Irene, his former landlady in La Crosse. Seeing her again meant fixing the radio, so he'd do whatever it took.

Monteverde worried about sparks from the radio igniting spilled fuel, so he insisted that the work occur outside the fuselage. With help from Harry Spencer and Clarence Wedel, Howarth moved several intact pieces of the radio to a small, hollowed-out igloo they dug in the snow under the left wing. They covered the boxes with cloth and pieces of parachute, then thawed them out with a battery-powered signal lamp. Shining the powerful light toward the radio warmed it enough to get the knobs turning and the insides unfrosted. For power, they connected the radio to the plane's batteries.

Knowing that the batteries wouldn't last, Wedel worked to fix the PN9E's gas-powered generator. He built a fire on the ice and placed the generator nearby. One by one, he thawed out its parts and dried its wires. Wedel then rebuilt it from scratch and somehow got it working. With a large quantity of fuel still in the B-17's tanks, they'd have power for as long as Wedel could keep the generator from breaking down or freezing. While it worked, he charged the batteries for the radio and scavenged lightbulbs from the cockpit. He wired the bulbs and strung them in the tail section, so no longer were the survivors condemned to darkness for more than eighteen hours a day.

Tucked in the little igloo under the wing, Howarth focused his attention on a long-range liaison radio, a transmitter and receiver that could communicate with ground stations or aircraft in flight. Depending on conditions, its signal could reach hundreds or even thousands of miles. For an antenna, it could use the skin of the airplane or the long wire from the Gibson Girl. Because the liaison radio was no longer attached to the plane, Howarth chose the wire, unspooling it along the ice.

His work settled into a punishing cycle: thaw the radio, remove his gloves, work until his fingers froze, warm his hands, then thaw the radio again. Often the cycle took a few minutes, and Howarth's hands cracked and bled. Between cycles he returned to the tail section, where his crewmates warmed his hands and encouraged him. Spina cringed as he heard Howarth crying from the pain in his fingers. Howarth knew that they were depending on him, and the pressure showed. Woody Puryear shuddered as he heard Howarth lament that the radio was too smashed and the rewiring too complicated.

"I can't do it, fellas," Howarth said. "I can't do it."

But as soon as his hands thawed enough to move his fingers, he returned to the igloo and resumed work. He studied torn and incomplete assembly diagrams and created jury-rigged replacements for broken parts. He worked around the clock.

After several exhausting days, Spencer and Wedel helped Howarth move the equipment to the PN9E navigator's compartment, an area behind the plane's broken Plexiglas nose. They blocked the opening with an inflated life raft, but it wasn't much warmer than the igloo. Every night the snow poured in, and every morning Howarth had to thaw out the radio and himself.

One night while Howarth worked, the crew heard what sounded like gunshots in their quarters: metal rivets began to pop as the tail section inched toward the widening crevasse underneath. With each unnatural movement of the broken B-17, the rivets that held together the fuselage panels protested by abandoning ship. To slow the slide into the crevasse, the crew gathered all available rope and parachute shroud lines. They lashed the tail section to the front half of the plane, twisting the lines to keep them taut. It was a temporary solution, but they hoped it would hold until rescue came.

On November 16, one week after the crash, Howarth announced that the new and old pieces of the rewired radio fit together. He flipped the power switch on his Frankenstein-like creation and the transmitter lit up. The receiver remained broken, so he didn't know if anyone heard him. But at least he could reach out farther for help than with the Gibson Girl. Howarth used Morse code to send SOS messages and to describe the PN9E crew's situation. Hoping that the receiver would soon work, too, he asked for replies with the MO, or magnetic orientation, from which his messages were originating. The crash had destroyed the B-17's direction finder; knowing the magnetic orientation might enable Howarth to send messages that included the plane's location.

Unknown to Howarth, his messages were heard as nearby as the U.S. Army's Bluie bases on Greenland, and as far away as a ham radio operator in Portland, Maine, some two thousand miles from the crash site, who relayed the SOS to military authorities. A message heard by an army radio operator in Greenland was garbled in spots, so the unidentified amateur in Maine filled in the blanks. One of Howarth's first messages read, in part, "Prep Negat Nine Easy [PN9E] crashed in glacier. . . . Have kept alive. Send help soon."

Howarth kept working on the stubborn receiver. Certain that he'd put it together correctly, he fretted over an instruction manual missing since the crash. The following day, after a nightlong blizzard, Clarence Wedel stepped outside the tail section, and there was the manual, uncovered by the blowing winds. Paul Spina and several other crewmen considered it a miracle.

Howarth pored over the manual and discovered that he'd incorrectly connected two wires. He switched them and made some adjustments, and then the receiver worked, too. He hailed Bluie West One. When the first faint reply came in, a frenzied Howarth was too excited to talk. Finally he yelled to his comrades, "We got 'em!"

The reply from Bluie West One promised supplies and help, either by plane or dogsled. It also instructed the crew to be on the lookout for a ship on the water nearby, and to shoot flares every evening at set times. Atmospheric conditions made it impossible to pinpoint the PN9E's magnetic orientation, but the description Howarth provided of being on a glacier northwest of a fjord was helpful.

In the hours and days that followed, Howarth continued to send and receive messages, each one proof of ongoing life in the wrecked B-17. Their rations had shrunk to two crackers and two pieces of cheese a day. But as engineer Al Tucciarone put it, with the radio working they felt like kings.

Monteverde captured the crew's feelings another way: if they survived this ordeal, they'd all owe their lives to Lolly Howarth.

10

FROZEN TEARS

NOVEMBER 1942

AT THE U.S. Army's Greenland bases, excitement about PN9E's radio messages contrasted with worrisome silence from McDowell's C-53. Hope ebbed about the fate of the cargo crew, though the five men wouldn't be presumed lost until they'd been missing for a month. In the meantime, efforts continued for both planes.

But again searchers were stymied by brutal weather. No flights left the Bluie bases on November 16 or 17. When the skies cleared on November 18, sixteen planes went aloft over the east coast, including the one that spotted Goodlet, Nash, and Weaver as the three Canadians walked toward the fjord. Storms and treacherous winds returned, and no planes left the bases from November 19 through 23.

On November 23, the PN9E crew exhausted their food supply, savoring their last few biscuits. They smoked their last cigarettes. Tucciarone cleaned out the ashtrays in the cockpit to give them a few final drags.

A lack of the most basic supplies came at a cost—literally. Ration boxes included toilet paper, but they used it up quickly. In its absence, the men dug into their wallets for the paper money they hadn't used for their Short Snorter memberships. They started with one-dollar bills, but soon those ran out and the men moved to larger denominations. The longer they spent on the ice, the more expensive personal hygiene became.

Monteverde and his crew fired several of their remaining flares and listened hopefully for the drone of an airplane engine, but each day of empty skies took a toll. Already weak and thin, their bodies burning muscle to stay warm, the crew's anticipation of impending rescue turned to anxiety. Maybe no one would spot them, and even if someone did, it might be impossible to navigate the crevasse field. When no planes arrived day after day, morale drained and anxiety descended into fear. Communicating with the Bluie bases on Howarth's radio would be worthless if all they could do was talk. They needed food and help before the cold picked them off one by one.

November 24 looked like another dispiriting day of lousy weather and no searches. Thanksgiving was two days away, and it seemed as though the men of the PN9E would spend it hungry, wet, shivering, and scared. But the headstrong commanding officer of the remote northern Greenland base called Bluie West Eight had other ideas. He commandeered a civilian passenger plane, filled it with supplies, rounded up a volunteer crew, and flew east into the storms toward the missing men.

For anyone else, it would have been reckless. For Bernt Balchen, it was routine.

BALCHEN (RHYMES WITH "walk in") was forty-three years into a remarkable life spent at the extreme edge of adventure. Powerfully muscled, with a square chin, thick blond hair, and Nordic good looks, he had the constitution of a draft horse and a fitting nickname: "The Last Viking."

The Norwegian-born son of a country doctor, Balchen joined the French Foreign Legion as a teenager, then transferred to the Norwegian Army as an artillery trainee. He was too late to fight in World War I, so he joined the Finnish cavalry under an assumed name to battle the Russian-supported socialists in the civil war of 1918. Left for dead on a battlefield when his horse was shot out from under him, Balchen recovered by relying on the strength and vigor that had made him a champion skier, cyclist, marksman, and boxer. He was not yet twenty.

After the war, while awaiting word on whether he'd been chosen to box for Norway in the Olympics, Balchen gave up his athletic career to become a pilot in the Norwegian Naval Air Force. Commissioned a lieutenant, in 1925 Balchen joined a rescue mission for famed Antarctic explorer Roald Amundsen, who had hoped to be the first man to fly over the North Pole. More polar adventures followed, as Balchen alternated between working with Amundsen and Amundsen's great rival, U.S. Navy commander Richard E. Byrd.

LEGENDARY AIRMAN BERNT BALCHEN. *(U.S. COAST GUARD PHOTOGRAPH.)*

With Byrd, Balchen piloted the plane that made the first airmail delivery from the United States to France. He took the controls during fierce storms and heavy fog over Paris, winning acclaim for saving the crew by setting down in the waters off Normandy. Already famous, Balchen gained worldwide renown in 1928 when he flew to remote northern Canada to rescue a German crew that had crashed after a transatlantic flight.

Balchen and Byrd had a contentious relationship, yet Byrd respected Balchen's flying skills and invited him to immigrate to the United States. In 1929, Balchen served as the chief pilot when a Byrd-led crew made the first flight over the South Pole. His exploits became so celebrated during the early years of flight that he found steady work as a test pilot for extreme weather conditions and as a consultant for aircraft makers and other fliers, including Amelia Earhart. In 1930 Congress passed a special act making him a dual citizen of Norway and the United States.

In an era of self-promoting aviators, some of whom were more skillful in press conferences than in the sky, Balchen was described as modest, even shy, despite a reputation among his peers as one of the best and bravest fliers of the age. Later in life, he'd shed his youthful reserve and display a streak of braggadocio.

Balchen kept busy throughout the 1930s, including a stint as chief pilot for a journey through Antarctica led by the American explorer Lincoln Ellsworth. He served as chief technical adviser for Norwegian Airlines from 1935 to 1940; as a cocreator of the Nordic Postal Union; and as a negotiator on an aviation treaty between the United States and Norway.

When Russia attacked Finland in 1939, Balchen was in Helsinki trying to obtain U.S. fighter airplanes for the Finnish Air Force. When Norway fell to the Germans the following year, he established the Norwegian Air Force training base known as Little Norway in Canada, to train pilots who'd escaped the Nazis. After handing that task to others, Balchen ferried bombers for the British. He also found time to indulge in his hobby as a watercolorist, favoring bold colors and the powerful scenery of Arctic vistas he knew and loved.

In 1941, an aide to U.S. Army general Henry Harley "Hap" Arnold met Balchen in the Philippines to ask him to command a secret air base in Greenland that would be used as a ferrying stop over the Atlantic. Soon Balchen was a colonel in the U.S. Army Air Corps and commander of its northernmost base, Bluie West Eight.

The Operation Bolero ferrying effort had barely begun when Balchen had a chance to demonstrate his rescue skills. In June 1942 a B-17 called *My Gal Sal* with a crew of thirteen ran out of fuel in bad weather and went down in Greenland. A radio operator at Bluie West Eight picked up the crew's distress calls and determined that the plane was about one hundred miles from the base. *My Gal Sal* was soon located from the air, but it wasn't clear how to rescue the crew. After dropping supplies, a search pilot noticed that melting ice had formed a temporary lake about sixteen miles from the downed B-17.

With Balchen aboard, a navy pilot named Dick Parunak landed an amphibious plane on the meltwater lake. Balchen hopped out and led a ground party on a two-day trek across soft ice, around open crevasses, and through glacial rivers. After reaching *My Gal Sal*, Balchen guided all thirteen members of the crew back to the rescue plane, and Parunak made two trips to return them all to Bluie West Eight.

Balchen received the Soldier's Medal, the army's highest award for heroism outside combat, and Parunak received the Distinguished Flying Cross. Two weeks later, the pair teamed up again to rescue Colonel Robert Wimsatt, commander of the U.S. Army's Greenland bases, and another flier whose plane went down on a patrol flight. After two successful missions and fifteen men rescued in two weeks, Balchen and Parunak jokingly named their army-navy partnership the Greenland Cooperative Salvage Company.

Now, four months later, Balchen faced a new test: guiding a commandeered C-54 Skymaster through late-autumn storms toward Koge Bay, hoping to add the nine men of the PN9E to his rescue total.

BALCHEN VIVIDLY DESCRIBED the challenges facing pilots in the far north: "When you fight in the Arctic, you fight on the Arctic's terms. . . . In the air you fight ice that overloads your wings and sends you out of control; you fight eccentric air currents over the ice cap that rack a plane and drop it several thousand feet without warning; you fight the fog. Most of the time you win, but sometimes you lose, and the Arctic shows no mercy to a loser."

A search flight he made for the PN9E nearly proved his point. "For twenty minutes our four-engine ship is tossed like a leaf in a cyclone. It's the severest turbulence I have ever encountered in an airplane, and I think in all my flying this is the narrowest escape of my life."

Hoping not to repeat the experience, on this day Balchen skirted around the storms. As he approached Koge Bay, the sky was getting dark. But before Balchen turned back, he saw a small red star rising in the distance. Then another. Then a third. He changed course to head toward the flares and dropped to a lower altitude. Looking down through the windshield, he saw the broken PN9E. He thought it resembled a crushed dragonfly on the ice.

He noticed how the bomber had snapped in two, and how the tail hung down at about a thirty-degree angle toward a gaping crevasse. Scouting the area beyond the wreck, Balchen concluded that the plane had gone down on the worst possible area of the ice cap: an active, crumbling glacier, surrounded by crevasses and deep ice canyons. The only approach on the ground, he thought, would be from the north, where the glacier seemed more stable, with fewer crevasses. Balchen considered it a miracle that anyone had survived the crash, not to mention the two weeks that followed.

Balchen noted the wrecked plane's position near Koge Bay, placing it around the intersection of longitude 65 degrees, 15 minutes north, and latitude 41 degrees, 18 minutes west. As he brought the big C-54 lower, Balchen ordered the crew to prepare bundles and crates of cargo they'd brought along, a cornucopia of food, fuel, two stoves, sleeping bags, clothing, and medical supplies.

WHEN BALCHEN'S PLANE approached the crash site, the men of the PN9E were clustered inside the tail section. At the sound of the four-engine Skymaster, all but O'Hara and Spina spilled outside onto the ice, tripping over each other like roaches startled by light. Monteverde and his men didn't believe their eyes until they saw packages falling from the plane at the end of parachute lines.

They'd been lost, but now they were found. Lolly Howarth's radio magic had led to their discovery. They huddled together for a prayer of thanks.

As he watched Balchen's plane circling overhead, Monteverde began to weep. He looked around and saw that his men were crying, too. Every last one. Tears rolled down their chapped faces and froze on their reddened cheeks.

ON THE FIRST three passes by Balchen's plane, fierce winds caught the cargo parachutes, turning them into sails and carrying the supplies into crevasses or far from the PN9E. One crewman grabbed a bundle only to be swept away by its wind-filled parachute. He let go just in time to avoid being pulled into a crevasse. Watching in horror, Balchen ordered the parachutes removed.

He made contact with Howarth by radio and told the men of the
PN9E to take cover inside their living quarters. Balchen flew a death-
defying fifty feet above the glacier and treated the next ten cargo drops
like pinpoint bombing runs. As he swept over the wreck, turbulent
air treated Balchen's ninety-four-foot-long plane the way a bull treats
a rodeo rider. Tucciarone felt certain that the C-54 and its crew would
be joining them on the ice. Yet the low-altitude cargo drops without
parachutes made a difference; several bundles even bounced off the
B-17's fuselage. Still, Tucciarone estimated that he and his crewmates
found only about one-fifth of the dropped cargo.

So eager were the men to gather the supplies that they ignored
the cold wind raking the ice cap. One after another they stumbled back into the tail section nearly blind, their eyelids frozen
together. The injured Spina and O'Hara blew warm air on their
crewmates' faces to thaw them.

Within a day, the men had recovered medical supplies and
five days' worth of food, mostly C rations, consisting of tin cans of
"wet" meat with hash or beans, and D-ration survival bars. Tea and
sugar had mixed together on the way down, so the men separated
the two using mosquito nets from the jungle kits. They found
two sleeping bags, which they gave to O'Hara and Spina. Both
men soon had company from crewmates, who took turns doubling
up with them for warmth. They recovered an Arctic Primus stove
that looked like an upturned blowtorch. But it was useless because
it wouldn't burn the leaded gasoline from the plane's tanks, and
they couldn't find any fuel dropped by Balchen.

A quart bottle of whiskey, dropped for "medicinal purposes,"
made the rounds, then reached Spina. A note in pencil on the label
said, "Take only in small quantity." Others heeded the advice, but
Spina belted down shot after shot. His compromised circulation
system experienced a powerful jolt. At first, his hands and feet tin-
gled; within ten minutes they throbbed with pain. Spina howled.
When the pain ebbed, he took the penciled instructions to heart.

Talking with Howarth by radio, Balchen offered a piece of advice that the PN9E crew, Harry Spencer in particular, could have used twelve days earlier: don't leave the wreckage unless everyone is roped together. Balchen reported that a dog team or motorsled team would soon be on the way. He also told them to watch the bay for a Coast Guard rescue ship, the cutter *Northland*. Before flying off, Balchen made plans with Howarth to stay in contact with twice-daily radio calls between the PN9E and Bluie West Eight.

On his way back to the base, Balchen tore a page from his diary and drew a sketch showing the location of the wreck, along with possible routes for an approach across the ice. He flew over Beach Head Station and dropped the weighted note for delivery to its commander, Max Demorest, the glacier scholar turned army lieutenant. Eleven days earlier, mechanical problems had forced Demorest to abandon the motorsled search for McDowell's lost C-53. With Balchen's map showing the site of the PN9E, Demorest would try another rescue, this time for the men of the B-17.

Although the frozen crewmen had food and basic supplies, time remained their enemy. Warmer weather during the two days following Balchen's flight caused the crevasse under the tail to open with a roar, reaching as wide as fifty feet across in spots. There was no suitable place for the men to hole up in the cramped front section, so they tightened the ropes anchoring the tail and hoped.

A more urgent worry was O'Hara's health. Gangrene gripped both feet. Monteverde fed him sulfa pills, but they made O'Hara delirious. The navigator talked gibberish and wouldn't eat. The worst soon passed, but his crewmates worried that O'Hara might die.

On November 26, Thanksgiving Day, two days after Balchen spotted them, Howarth sent the PN9E crew's most urgent call for help. With O'Hara's deteriorating condition preying on everyone's mind, he tapped out in code:

"Situation grave. A very sick man. Hurry."

AFTER RESCUING THE Canadians, the *Northland* had been ordered to cruise through the thickening ice toward the presumed
vicinity of McDowell's C-53 and Monteverde's B-17. Now, with
Balchen's discovery, the priority became the B-17.

The Greenland Patrol commander, Rear Admiral Edward "Iceberg" Smith, told the *Northland*'s captain, Lieutenant Commander
Frank Pollard, to head toward Koge Bay, as close as possible to the
PN9E location described by Balchen. Smith had concerns about
sending the ship into the ice-filled waters so late in the season,
but he understood that Monteverde's crew needed the *Northland*'s
help. Because ice already clogged parts of Koge Bay, the *Northland* would anchor to the east, in the body of water known to the
Americans as Comanche Bay, named for a Coast Guard cutter by
that name. Once there, the *Northland* would be close to Beach
Head Station and about thirty miles from the wrecked PN9E.

THE MOTORSLED TEAM of Lieutenant Max Demorest and Sergeant Don Tetley also joined the rescue effort. Their attempts to
find McDowell's C-53 had failed when their sleds conked out,
but the sleds were back in working order, and now they were
headed toward a known destination. Joining them was a dogsled
team led by Johan Johansen, a Norwegian fur trapper and survival expert who'd been stuck in Greenland since the Nazis had
invaded his homeland. Johansen had become a civilian adviser to
the U.S. Army, a job that distracted him from his worries about
his wife and son in occupied Norway.

The evolving plan was modeled on the rescue of the three Canadians, with a few wrinkles. On paper, it called for the motorsled
and dog teams to travel over the ice cap on the inland side of the
rugged coastline and mountains. They'd go from Beach Head Station to Ice Cap Station, then approach the PN9E from the northeast to avoid the worst crevasse fields. After reaching the bomber,

they'd lead or carry Monteverde's crew back the same way to
Beach Head Station, where the *Northland* would be anchored just
offshore in Comanche Bay. The PN9E crew would be taken by
motorboat to the waiting ship.

If everything went smoothly, all nine men would be safely
aboard the *Northland* in three or four days. The plan sounded so
straightforward that it was easy to imagine Monteverde's crew
sharing stories and toasting their good fortune alongside Goodlet,
Nash, and Weaver in the ship's sick bay.

But as the events of November 1942 had already proved, almost nothing goes as planned on Greenland's ice cap.

No one understood that better than Balchen, who once wrote,
"The Arctic is an unscrupulous enemy. It fights with any weapon
that comes to hand, it strikes without warning, and it hits hardest
just when you think the fight is won."

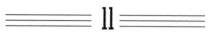

11

"DON'T TRY IT"

NOVEMBER 1942

As THE *NORTHLAND* fought through gales toward Comanche Bay, the rear admiral at the helm of the Greenland Patrol, Edward "Iceberg" Smith, expressed second thoughts about its role in the rescue mission. Smith wanted as much as anyone to save the PN9E crew. But he didn't want to lose 130 Coast Guardsmen and a vital ship under his command.

Smith's PhD in oceanography and the quarter century he'd spent in icy waters made him one of the world's foremost authorities on the awful conditions facing Lieutenant Commander Pollard and the men of the *Northland*. Although the ship had a reinforced hull and was built for Arctic operations, it wasn't an icebreaker. That meant ice could be a ship-breaker. Calamity could come swiftly, from an iceberg, or slowly, if the *Northland* became trapped in Greenland's coastal ice pack. In that case, the ship would be lost to the war effort for months, and the crushing pressure might leave the *Northland* crippled or worse. Smith knew the *Northland*'s strengths and limits firsthand: he'd been its commander for more than a year.

One of the Greenland Patrol's converted fishing boats, the *Ak-lak*, had already abandoned the search and left the coast after its anxious captain told Smith that "further delay will seriously endanger ship and personnel." Smith's radio messages to the *Northland* urged Pollard to consider the same prudent course.

At first, Smith trod lightly, respecting Pollard's prerogatives as the ship's captain. The admiral issued an inquiry rather than an order: "In view of lateness of [the] season and relative risks involved, and latest information you may have received from Ice Cap Station," he radioed the *Northland*, "do you consider further operations advisable?" It was easy to read Smith's message as offering an exit strategy, a way for Pollard to pull away from the rescue effort without losing face.

The response couldn't have been clearer, and it couldn't have comforted Smith. "Extremely hazardous . . . five miles south of Comanche Bay entrance . . . in rafted pack ice thickly interspersed with floe bergs, growlers around us . . . visibility poor." But then, to answer Smith's question, the ship radioed flatly: "Continuing attempts to work free and make base at Comanche Bay."

Smith didn't relent. Perhaps, he said in a follow-up message, the *Northland*'s doctor should go ashore to spend the winter at Beach Head Station. The injured PN9E crewmen could be brought to him there, allowing the *Northland* to head toward safer waters. The ship could return and fetch everyone from Beach Head Station in the spring. Smith still wasn't issuing an order, but he punctuated the suggestion with a warning: "Do not take risks this late in season."

Again Pollard demurred: "Do not, repeat not, deem it advisable for *Northland* medical officer to proceed with rescue party, and will keep him aboard ship unless otherwise directed." In other words, unless Smith gave a direct order, the *Northland* and its crew would remain on the job. Pollard's reply also informed Smith that the *Northland* had carved through the ice and had nearly reached

Comanche Bay. The message ended with an ominous weather report: "Fog."

Pollard and Smith were locked in a respectful standoff, and both men knew that Pollard would prevail barring a dramatic change in circumstances. Central to the Coast Guard culture was a belief in testing every reasonable limit to complete a rescue. A decade earlier, in the 1930s, the service's highest-ranking officer captured that outlook in a Coast Guard creed built around the phrase "I shall sell life dearly to an enemy of my country, but give it freely to rescue those in peril." Pollard couldn't bear to turn away from Howarth's plaintive message: "Situation grave. A very sick man. Hurry."

Smith might have taken a harder edge with Pollard had he known that a new rescue plan was brewing aboard the *Northland*. Despite repeated back-and-forth messages between the two, the captain of the *Northland* curiously, perhaps purposefully, failed to mention that the approach used to rescue the Canadians might not apply, after all, with the American crew.

To begin with, reaching the coastline on foot or by motorsled would be far harder for Monteverde's nine-man crew than it had been for Goodlet, Nash, and Weaver, who were exhausted but still able-bodied. In a follow-up message from Howarth in the PN9E, he described the "very sick man"—Bill O'Hara—as having "frozen feet, a touch of gangrene, high fever." A further complication was the instability of the Koge Bay glacier. Between the wrecked PN9E and the coast were untold hidden crevasses, one of which had nearly killed Harry Spencer.

In fact, the terrain proved too much for expert dogsledder Johan Johansen and his team. They turned back, defeated by deep, finely powdered snow and the tall waves of crusted snow called *sastrugi*. Even if Lieutenant Max Demorest and his motorsled team could reach the PN9E, there was no guarantee they could get the survivors to Beach Head Station, perhaps forty miles away, depending on the route. In addition, with fog and storms approaching, the *Northland*'s motorboat might struggle or capsize in Comanche Bay.

With doubts rising about how best to reach the PN9E crew, Lieutenant John Pritchard, the *Northland*'s Duck pilot, dreamed up a daring new plan that would eliminate the need for motorsleds and motorboats altogether. It also would remove arduous travel by sea, and would minimize the risk of crevasses. Pritchard proposed flying the Duck to the ice cap, landing briefly, leading the B-17 crewmen to his little plane a few at a time, then flying back to the *Northland*. Then he'd do it at least twice more to get all nine men off the ice.

JOHN PRITCHARD (LEFT) ON THE DECK OF THE *NORTHLAND* AS THE CHIEF BOAT-SWAIN'S MATE INSPECTS THE DUCK. *(U.S. COAST GUARD PHOTOGRAPH.)*

It was likely the same plan Pritchard had hoped to use during the search for Goodlet, Nash, and Weaver, but he'd been unable to find them. Now, with Balchen's coordinates, he knew exactly where to locate Monteverde's crew, so it made perfect sense that Pritchard and the Duck's radioman, Benjamin Bottoms, would be eager to show what they and their Duck could do.

PRITCHARD'S PLAN WAS elegant, efficient, brave, and dangerous. It also was unprecedented. No plane had ever landed on Greenland's ice cap and then taken off again. On each of three or more round-trips, Pritchard would have to contend with buffeting winds, blowing snow, and treacherous haze, the same conditions that contributed to the PN9E's crash and numerous others. Also, he'd have to avoid hidden crevasses located between his landing spot and the broken bomber.

Pritchard's plan to save the B-17 crew was bold, but it wasn't reckless. It fit the selfless oath he'd taken as a Coast Guardsman. It's also easy to imagine another, more personal motive: Pritchard's younger brother Gil was a B-17 pilot, flying combat missions in a Flying Fortress over North Africa.

By neglecting to inform Rear Admiral "Iceberg" Smith of the emerging Duck-centric rescue plan, *Northland* captain Frank Pollard might have been operating on the theory that it's better to ask forgiveness than permission. Another possibility was that Pollard hadn't yet decided to allow it.

Later, Pollard would explain that Pritchard had been determined to land on the ice cap in the Duck as soon as Balchen located the PN9E. Pollard said that he was persuaded to approve the mission by what he called "the touching appeal that was contained in the simple messages being sent out by the wrecked crew on their hand-powered radio set. They kept repeating they were getting weaker and told of two men suffering from advanced cases of gangrene, of other injuries and hardships." It wasn't clear to whom

Pollard was referring when he mentioned a second man with gangrene, but he apparently meant Paul Spina.

With Admiral Smith expressing qualms about the *Northland*'s role to begin with, Pollard wouldn't test his commanding officer's resolve. He'd tell the head of the Greenland Patrol what was happening only *after* the Duck took flight.

BY THE TIME he volunteered for the PN9E rescue, John Pritchard had gone through hell and humiliation to prove himself a gifted, fearless pilot.

Born in January 1914 in Redfield, South Dakota, Pritchard was the eldest of five children. Gil, his bomber pilot brother, was a year younger, followed by two more boys, one of whom died in infancy. Last was a girl, their baby sister Nancy. Their mother, Virginia, ran a strict home while working as a children's book reviewer. Their father, John Pritchard Sr., was a cattleman and a banker, but he lost everything when a late-spring storm in 1926 wiped out his herd. Reduced to selling applesauce, John Senior moved the family to Los Angeles for a fresh start.

In Nancy's eyes, her brother John was their parents' favorite. In the tradition of firstborn sons, John Junior was a responsible, dependable boy. The family's German nanny proclaimed that someday he'd be president. John cruised through Beverly Hills High School with a mix of A's and B's, and worked as a paper boy for the *Los Angeles Times*, where his father had found work as a circulation manager.

From age twelve, Pritchard dreamed of becoming a naval officer, but he couldn't collect the required recommendations for the U.S. Naval Academy. Instead, after high school he joined the navy as an enlisted man, then spent two and a half years trying to blaze a path to a career as an officer. He served nine months at sea and endured a hernia operation, and eventually he gathered enough support for admission to Annapolis. Pritchard passed all the tests—except for one, in geometry, where he fell four-tenths of a point below the bar. He'd reached the age limit of twenty for admission, so retaking the test the following year wasn't an option.

Pritchard swallowed his disappointment and set his sights on the U.S. Coast Guard Academy, where he'd be eligible for admission until he turned twenty-two. But again he hit a snag. He ranked eighty-eighth out of one thousand applicants on the Coast Guard exams, but only fifty-eight cadets were offered spots. Pritchard tried again the following year, but he was foiled by the physical exam: a blood test claimed that he had syphilis.

Watching their son's dreams and reputation unravel, Pritchard's parents mounted a feverish letter-writing campaign to powerful men from California to Washington. John Pritchard Sr. won support from a U.S. senator and a congressman. Virginia Pritchard wrote an impassioned letter to an old friend who'd become administrator of the U.S. Agricultural Adjustment Administration, who forwarded her letter to the secretary of the treasury. When campaigning for her son, Virginia Pritchard bared her political soul, declaring her love for the New Deal and her FDR-inspired conversion to the Democratic Party after a lifetime as a Republican. "We have no political friends," she pleaded. "If in any way you can help this son of ours, we shall be more than grateful."

John Pritchard took a second, more reliable blood test for syphilis, which came back negative. Yet still he was out of luck: the cadet class was full. His parents kept up the pressure, and their efforts paid off. Six days before the start of the school year, an accepted cadet dropped out and John Pritchard proudly took his place in the Coast Guard Academy's Class of 1938. His roommate, a future vice admiral named Thomas Sargent, called Pritchard "the happiest man I have ever known."

"At reveille," Sargent recalled, "he would practically jump out of his bunk and, in spite of rain, snow or darkness, he would say, 'Good morning, Tom, what a great day' and break out in song.

He had a good singing voice, and his favorite rendition was 'The Grandfather's Clock'—he knew all the verses. At first, starting the day like this was a little wearing, but his enthusiasm for life was so infectious I actually looked forward to reveille."

Upon graduation, the blue-eyed, brown-haired Ensign Pritchard stood five feet, ten inches tall, and weighed 145 pounds. He carried his thin frame at attention, shoulders back, a posture that his sister said reflected self-assurance, not cockiness. A good-looking man, Pritchard had an oval face he arranged in a thoughtful expression. He had several girlfriends, but was in no rush to marry. He seemed more interested in building his career while looking out for his sister and his friends. During one of his

JOHN PRITCHARD'S COAST GUARD ACADEMY PORTRAIT. *(U.S. COAST GUARD PHOTO.)*

first postings, aboard a Coast Guard cutter in the Bering Strait, Pritchard became best friends with Ensign Harry "Tick" Morgan. He decided that Morgan would make an ideal match for Nancy. "He said, 'I'm saving Nancy for Tick and Tick for Nancy,'" she recalled.

After Alaska, Pritchard was accepted for flight training and became Coast Guard Aviator No. 82. Promoted to lieutenant, he served at the Miami Air Station until February 1942. Nancy spent six weeks of her summer vacation from college visiting him in Florida. During that time, Tick Morgan's ship came in, and so did Nancy's. They wed two years later and stayed married for the next sixty years.

After Miami, Pritchard was assigned to the *Northland*, as pilot of the ship's Duck. Except for a brief period back in Florida, he spent most of 1942 flying the little biplane countless miles over and around Greenland.

THE MAN WHO sat behind Pritchard in the Duck was the *Northland*'s ruddy-cheeked radioman first class, Benjamin A. Bottoms. A year older than Pritchard, the bearded, blue-eyed Bottoms could have shared a wardrobe with his pilot. He, too, stood just over five-feet-ten and weighed 145 pounds. By coincidence, Bottoms also had a sister named Nancy, who was his twin.

A farm boy from Cumming, Georgia, Bottoms enlisted in the Coast Guard after graduating from high school. He spent a decade moving from station to station, ship to ship, reenlisting three times and picking up a nickname fitting his Deep South accent: "Georgia Cracker." In 1937 he married Olga Rogers, a fisherman's daughter from Gloucester, Massachusetts. They settled with her son Edward, whom Bottoms called "Bud," near the Coast Guard Air Station in Salem, Massachusetts.

In December 1939, Bottoms found himself in an unusual predicament for a Coast Guardsman: adrift at sea. He and three other

servicemen were aboard an amphibious plane called a Douglas
Dolphin when it was forced down in fog twelve miles off the Massachusetts coast. Damaged in the landing, in danger of capsizing,
the plane drifted for twenty-five miles and was battered by waves
for more than a day before being towed to safety by a Coast Guard
cutter. No one was injured, but Bottoms would never forget how
it felt to need rescuing.

He served for five months on the *Northland* in 1941, then returned to Massachusetts with measles. When he recovered, Bottoms could have escaped the demanding Greenland duty. Instead
he volunteered to return as the Duck's radioman. He rejoined the
ship in February 1942, around the same time as Pritchard, and the
two became a team. Bottoms impressed his crewmates and superiors with his skills and work ethic. Weeks before the PN9E rescue
mission he was recommended for promotion to chief petty officer.

RADIOMAN BENJAMIN BOTTOMS BEFORE HE GREW A BEARD.
(U.S. COAST GUARD PHOTOGRAPH.)

ON THE COLD morning of Saturday, November 28, 1942, the *Northland* pushed through the ice into Comanche Bay. The ship's communication officers picked up Howarth's faint distress calls, which gave Pritchard and Bottoms a bearing they could follow to the PN9E crash site. The ship's crew lined the rail as the little plane taxied away. Less than twenty minutes after the *Northland*'s anchor splashed into the bay, the Duck was in flight.

As Pritchard and Bottoms flew toward the downed B-17, Colonel Bernt Balchen and his crew were simultaneously flying over the crash site in their borrowed C-54 Skymaster. Balchen dropped more medical supplies for O'Hara's gangrenous feet, along with extra sleeping bags, canned chicken, sausages, soups, and candy. By radio, Balchen told Lolly Howarth that he'd spotted two mo-

THE GRUMMAN DUCK, PILOTED BY JOHN PRITCHARD, WITH BENJAMIN BOTTOMS SERV-ING AS RADIOMAN, TAXIS AWAY FROM THE *NORTHLAND* ON NOVEMBER 28, 1942. *(U.S. COAST GUARD PHOTOGRAPH BY CHARLES DORIAN.)*

torsleds carrying Lieutenant Max Demorest and Sergeant Don Tetley some twenty miles away. They were making good time across the ice, each one towing a cargo sled loaded with equipment and supplies. Balchen expected the sledders to arrive at the wreck late that night. He told Howarth to shoot flares after eight o'clock to guide them in.

As Balchen circled overhead, he received a radio call from Pritchard in the Duck. The two pilots knew one another from Pritchard's occasional landings at Balchen's base. Balchen told him about a level area relatively free of crevasses not far from the PN9E. There, he said, Pritchard could make a wheels-up landing on the Duck's belly, using the central pontoon like a sled on the ice. Balchen flew over the potential landing site and dropped one hundred feet of rope, snowshoes, and bamboo poles to help Pritchard and Bottoms trek to the wreck after they landed.

Running low on fuel, Balchen turned the C-54 back across the island toward Bluie West Eight, confident that Pritchard and Bottoms in the Duck, and Demorest and Tetley on motorsleds, had the rescue of the PN9E well in hand.

PRITCHARD CIRCLED THE Duck above the bomber, then leaned forward on the control stick and brought his plane low over the PN9E. Pritchard buzzed so close to the ground that several of the icebound men took cover inside the tail section. Pritchard tossed out a small can with a red flag tied to it. Inside, Monteverde's crew found a note that Pritchard had written while aboard the *Northland*. It inquired about ground conditions and asked Monteverde's crew to stand on the B-17's right wingtip if they thought it was safe for the Duck to land with its wheels down. If they thought that he should land wheels-up, on the Duck's belly, they should stand on the left wing. If landing was too dangerous to consider altogether, they should gather on the PN9E's tail. As a signoff, the note read, "If there's a 60-40 chance, I'll take it."

When Monteverde read that line aloud, several men wiped away tears.

The desperate bomber crewmen looked at one another. They feared that the crevasses would swallow the Duck like an alligator opening its jaws under a mallard. Without a word, they climbed atop the tail to tell Pritchard and Bottoms that it was too danger-ous to land. In doing so, they knew that they were waving off what might be their best chance for survival.

Minutes later, Howarth and Monteverde reached Ben Bottoms by radio, and Monteverde reinforced the message. He told the Duck's crew, "Don't try it."

Pritchard buzzed them again, wagging his wings and waving from the cockpit. The dejected PN9E crew thought that was the last they'd see of the plane. But as they watched, Pritchard circled lower and lower over an area several miles away. On Pritchard's orders, Bottoms radioed: "Coming in anyway."

A member of the PN9E crew muttered, "He won't make it, poor fellow." Several couldn't bear to watch. They climbed back inside their shelter.

Adding to the danger, Pritchard disregarded Balchen's advice about the best way to touch down. He hand-cranked the Duck's landing gear into place, intending to treat the ice cap the same as he would a paved tarmac. It was a calculated risk. A belly-down landing might damage the Duck's fuselage or curl its propeller, rendering it yet another squished bug on the ice cap. On the other hand, a wheels-down landing could lead to the same result. Five months earlier, when the pilot of the first plane from the Lost Squadron tried a wheels-down landing, he'd flipped his fighter onto its back. But Pritchard went with his gut: wheels down.

Pritchard clicked through his eight-point landing checklist: pro-peller set at low pitch, cabin hood locked open, tail wheel locked, and so on. He pushed forward on the control stick, and the Duck met the ice cap. The wheels touched, then sank into deep snow. The plane's bulbous nose seemed intent on burrowing downward into a catastrophic somersault. Pritchard fought back, relying on the big central pontoon and the wing floats to keep the plane level. Several times the tail lifted, threatening to cartwheel over the nose and destroy the plane and both men aboard. Pritchard kept fighting.

He brought the Duck to a stop, completing the first planned, successful, wheels-down landing on a Greenland glacier.

To AVOID THE web of crevasses, Pritchard had landed about two miles from the wreck, far from where Balchen dropped the rope, snowshoes, and bamboo poles. Pritchard and Bottoms climbed out of the Duck equipped with little more than a broomstick to test for hidden ice bridges.

For more than an hour, the two Coast Guardsmen shuffled, poked, trudged, plowed, and slid across the glacier. At one point Pritchard slipped into the mouth of a crevasse, but he managed to catch an edge and pull himself out. When they reached the wreck, Pritchard approached each crewman of the PN9E with an outstretched hand.

Monteverde told Pritchard, pilot to pilot, "You shouldn't have landed. Now you may not be able to get off."

"I came prepared to stay," Pritchard answered.

It was a good line, but in fact Pritchard intended to take off as soon as possible. He told Monteverde that the Duck would take two men immediately and would return the next day for the rest. O'Hara and Spina were the worst off, and Monteverde wanted them to go first. But the navigator's frozen feet and the engineer's broken arm and other injuries made it impossible for either to reach the Duck without being carried on sleds. They'd have to wait until the Duck returned with hand sleds or they could be carried on Demorest and Tetley's motorsleds. Pritchard suggested that Monteverde's frostbite made him a good candidate. But Monteverde was captain of the PN9E. He'd be the last to leave.

Monteverde chose two of the best-liked men aboard the bomber, Al Tucciarone and Lloyd "Woody" Puryear. Both needed medical care, Puryear for feet that had frozen in his leather boots, and Tucciarone for the broken ribs he'd suffered in the crash. Both were weak with hunger and chilled to the bone, yet Monteverde considered them strong enough to reach the Duck on foot.

The timing was providential for Tucciarone, who'd begun having visions. He felt certain that he'd seen a giant image of Jesus Christ in the Greenland sky. Puryear felt fragile and stiff from the cold, and his mind had become so foggy from hypothermia that he barely registered that he was leaving his new friends. When it dawned on them that they'd been chosen, Puryear and Tucciarone both declined. They argued that copilot Harry Spencer and volunteer spotter Clint Best should take their places. Spencer and Best refused, Monteverde pulled rank, and that was that.

Despite his fall into the crevasse, the resilient Spencer felt the strongest of the PN9E crewmen. He volunteered to make the trek to the Duck, in case Tucciarone or Puryear needed help along the way. Pritchard roped them all together, separated at thirty-foot intervals, and led them onto the trail that he and Bottoms had made on the way in.

Nineteen days of vicious cold and little food had sapped the young airmen's strength more than they'd realized. Puryear could hardly lift his feet, and Pritchard and Bottoms dragged both him and Tucciarone almost as much as they walked. Spencer helped, but he couldn't do much. Soon both Puryear and Tucciarone fell face-first into the snow from exhaustion. Puryear felt himself surrendering to sleep, a precursor to death on the ice cap. Bottoms propped up Puryear, put a cigarette in his mouth, and lit it for him. He encouraged the young man from Kentucky onward, half carrying him much of the way. Pritchard tried to raise Tucciarone's spirits by joking that Tucciarone's mother would have homemade spaghetti waiting for him.

Fearing the approach of dusk, Pritchard and Spencer hurried ahead to prepare the Duck for takeoff. Pritchard decided to crank up the wheels, so they wouldn't drag through the snow. But in the four hours since landing, the Duck's floats had frozen to the glacier. Pritchard and Spencer dug out the wheels and rocked the little plane to break it free of the ice. They turned it in the opposite direction, using the ice as a pivot, so Pritchard could fly out the way he'd come in. By the time Bottoms and the two zombie-like PN9E crewmen reached the Duck, it was ready to go.

Puryear and Tucciarone half climbed and were half carried into the empty compartment below the Duck's cockpit. Pritchard and Bottoms strapped themselves in and prepared to leave. Pilots seek to take off into a headwind because it's easier to generate lift, but Pritchard didn't have that option. Spencer gave the plane a push to help it on its way, then watched as the Duck gained speed for a difficult and dangerous downwind takeoff.

Tucciarone and Puryear couldn't see what was happening from inside the cargo compartment. They clasped hands and prayed as they felt the Duck rise, then fall, their stomachs and their hopes doing the same. The plane bounced from frozen hill to frozen hill, icy hummock to icy hummock. The span between impacts became longer and longer. The plane rose higher. The Duck was airborne. Tucciarone heard Pritchard and Bottoms scream for joy. He and Puryear did the same. Pritchard had made history a second time, as the first pilot to land *and* take off from the Greenland ice cap.

Before Pritchard began the brief flight back to the *Northland*, he flew over the wrecked bomber and again waggled his wings.

The sun was setting, so the *Northland* shone its big searchlights onto the water of Comanche Bay to help Pritchard gauge his altitude on approach. Twenty minutes after takeoff from the glacier, he came down as smoothly as if he'd been planning a picnic on a midwestern lake. The ship's crew lined the rail again, cheering as Pritchard taxied the Duck to its side.

Five hours after watching the Duck fly off, the *Northland* crew hoisted the plane back onto the deck. Gingerly, they pulled the two rescued men out of the Duck's belly. Ragged, thin, and frostbitten, but safe.

Lieutenant Commander Pollard fired off a celebratory message to "Iceberg" Smith reporting that Al Tucciarone and Woody Puryear were onboard, and that the *Northland*'s Duck was ready and willing to return for more. Smith raised no protest, so Pollard elaborated in a second message, explaining the plan for the first time to his commanding officer. In the process, he torpedoed the idea of relying on motorsledders Max Demorest and Don Tetley for the remaining rescues:

If weather permits, will evacuate seven B-17 personnel via *Northland* plane in two flights commencing daylight tomorrow, Sunday. All personnel excepting one are in critically weakened

AL TUCCIARONE IS CARRIED DOWN FROM THE DUCK AFTER THE PLANE'S RETURN TO THE *NORTHLAND* ON NOVEMBER 28, 1942. *(U.S. COAST GUARD PHOTOGRAPH BY CHARLES DORIAN.)*

condition. Several serious cases [of] gangrene and one broken arm. . . . Will have two ski/hand sledges constructed within few hours for use in transporting B-17 personnel to *Northland* plane. B-17 men are not equal to ordeal of 50-mile sledge trip, which may entail being on trail and in Ice Cap Station for two or three more weeks without urgently needed medical attention and hospitalization. Believe present favorable weather will not last beyond tomorrow. Evacuation by plane deemed imperative to prevent deaths and also losses of limbs and extremities from progressive gangrene. Medical officer urges expeditious hospitalization ashore for all B-17 personnel. Two men brought aboard today by *Northland* plane have grave debility. Puryear also has gangrenous toes and Tucciarone breathes with difficulty because of fractured ribs.

Smith accepted the rescue plan without objection, asking only that Pritchard ensure the destruction of the PN9E's Norden Bombsight, along with any other equipment or documents that would help the enemy. Before signing off, Smith asked Pollard to tell the three Canadian crewmen still aboard the *Northland* that their families had been notified that they were safe.

When they climbed down from the Duck, Pritchard and Bottoms accepted congratulations and hearty slaps on the back. They secured their plane for the night, ate, then went to their bunks to prepare for a big day ahead.

Tucciarone and Puryear were carried to sick bay, where the ship's doctor waved away piles of food and drink offered to them by the *Northland*'s crew. The doctor prescribed a strict diet of hot soup, hotter coffee, and most of all, rest.

Before going to sleep, Tucciarone and Puryear had just one request: John Pritchard's autograph. With that in hand, the two survivors and the pilot who rescued them would be Short Snorters together.

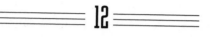

12

"MOs—QUICK!"

NOVEMBER 1942

AFTER THE DUCK returned to the *Northland*, the night on the
ice cap was clear and the weather tame. It was tempting to imagine that Greenland had surrendered. The seven remaining PN9E
crewmen bundled together in the bomber's tail section, warmed
by the thought that two of their friends were safe and their own
imprisonment was nearly over. As darkness fell, they fired bright
red flares into the starlit sky. The official purpose was to guide
the motorsled drivers to the wreck, but the flares also seemed a
fitting celebration of the anticipated end of their ordeal.

Seeing the red starbursts, Max Demorest and Don Tetley knew
they were close. In fact, the lights on their sleds were already visible from the PN9E. They shot off a flare of their own to let the
bomber crew know they were within range. But a mile or so from
the plane, the motorsled drivers halted at a menacing sight: their
intended route was across an active glacier beset by crevasses. Demorest, the Princeton PhD glacier expert, and Tetley, the Texas
horseman turned Arctic explorer, climbed off their motorsleds to
assess the danger. Agreeing that it was too perilous to proceed on

their sleds at night, they strapped on skis and grabbed poles to test the way. They'd find a safe trail through the maze of crevasses on foot, and then return to the sleds to drive along that footpath.

Carrying a bundle of supplies and threading their way around the ice chasms, Demorest and Tetley reached the PN9E after midnight. They found Monteverde's crew awake and in high spirits. After hearty greetings, the seven remaining survivors broke out a can of chicken that they'd saved for just such an occasion. Demorest dressed O'Hara's gangrenous feet and treated him with antibacterial sulfa drugs and morphine for the pain. Tetley described O'Hara as suffering from "frozen feet and body poison." The diagnosis reflected his belief that the navigator's shriveled feet had become infected and that immediate evacuation was essential for any chance to avoid amputation.

After treating O'Hara, Demorest checked Spina's arm to ensure that the bones were knitting properly, then applied fresh bandages. Then he went man to man, wrapping frostbitten fingers and toes. As the ministrations continued, the men of the PN9E told Demorest and Tetley about one further indignity: a steady diet of dry rations, a lack of fruit or vegetables, and little exercise had left most of them constipated.

When the medical work was complete, Demorest and Tetley shared news of the war. They regaled the PN9E crew with the triumph of Operation Torch, the joint British-American invasion of North Africa. They also reported the bad news that McDowell's C-53 still hadn't been found. Then they explained their plan to take everyone to Ice Cap Station and then on to Beach Head Station, on foot or motorsled, if the Duck didn't return.

Even with Tucciarone and Puryear gone, the bomber's tail section felt too cramped for Demorest and Tetley to spend the night. Also, they worried that a blizzard might cover their sleds while they slept, delaying them in the morning. They told the bomber crew that they'd sleep in tents by their motorsleds and return around 10:00 a.m. The PN9E crew urged them to stay, fearful of the motorsled men walking in the dark amid the crevasses. But Demorest and Tetley wouldn't hear of it. As the party broke up, Monteverde made plans to greet Demorest and Tetley with a hot breakfast.

After the motorsled men left, the PN9E crewmen were too excited to sleep. They talked about getting back home and agreed how nice it would be to sleep on a cot again. Monteverde reminisced about the warmth of California, and anticipated a spaghetti feast that Tucciarone had promised everyone. Hearing that, Spencer cranked open a can of spaghetti that Tetley had left behind. He warmed it over the fire and gave his crewmates one forkful each. Spencer joked that he wanted all of them to taste spaghetti before Tucciarone did.

At O'Hara's suggestion, they prayed together before turning in. Afterward everyone lay in silence, but Spina knew that no one was asleep.

THE FOLLOWING DAY, November 29, 1942, Winston Churchill delivered a worldwide radio address declaring a turning point in the war, the Allies' first "bright gleam of victory." He hailed the defeat of German field marshal Erwin Rommel in Egypt, and the American-led landings in Algeria and Morocco. Despite the good news, for the first time since Pearl Harbor the war was overshadowed in the United States by a homegrown tragedy: nearly five hundred dead, many of them servicemen, in the Cocoanut Grove nightclub fire in Boston.

Aboard the Coast Guard cutter *Northland* in Comanche Bay, neither Churchill's speech nor the nightclub fire was the big news. Thoughts and hopes were focused on the PN9E crew, and on Pritchard, Bottoms, and the Duck.

When daylight broke, wispy fog and steely clouds replaced the clear sky of the previous night. Pritchard checked the weather reports and saw that snow was on the way and the overcast was thickening. At 8:00 a.m., visibility was approaching twenty miles. But it soon deteriorated, falling to perhaps half that distance an hour later. The fog was getting murkier, the snow heavier, the sky darker. By noon, visibility would be less than one mile. If they chanced it, Pritchard and Bottoms would be flying by the seat of their pants.

But Pritchard believed that he, his radioman, and the Duck could handle the weather, and it wasn't likely to improve the next day or the day after. During the previous nine months, he'd become expert in the tricky flying that was required over and around the island. He knew how to hold steady through wild bursts of turbulence, loss of visibility, and dangerous haze that blurred the horizon. Pritchard took those risks seriously, but he'd also seen firsthand how the men of the PN9E were suffering. This was their twentieth day on the ice, and they'd waited long enough. Pollard, the *Northland*'s captain, trusted Pritchard's judgment and allowed him to proceed.

The decision was sealed when Lolly Howarth awoke that morning and radioed the *Northland* that it was a beautiful day on the glacier, with good visibility.

As the ship's crew again lined the rail, Pritchard and Bottoms gained speed along their water runway and took off at 9:29 a.m. The flight normally took no more than twenty minutes, so Pritchard and Bottoms could expect to join Demorest, Tetley, and the remaining seven PN9E men for their scheduled ten o'clock breakfast.

The flight over Koge Bay was uneventful. As Pritchard flew above the wreck, Bottoms tossed out the stretcher-sleds built by the *Northland*'s carpenters, so they'd be near the bomber to transport O'Hara and Spina to the Duck. At the sound of the plane, Harry Spencer emerged from the B-17's tail to retrieve the sleds. Monteverde intended to have Demorest and Tetley use their motorsleds to carry O'Hara and Spina to the Duck, but Spencer decided to assemble the handmade stretcher-sleds nonetheless. Spencer also made a fire to cook breakfast for the rescuers and his fellow crewmen, but the PN9E survivors wanted only coffee. They were saving room for an expected banquet aboard the *Northland*.

As he worked, Spencer took several breaks to slip inside the tail and give updates to Spina and O'Hara, who remained in their sleeping bags. On one visit, he said, "It won't be long now. The sleds are almost here and the plane is due any minute. I guess we can kiss the Ice Cap goodbye." Several other men came inside to pat O'Hara and Spina on the back for good luck.

WHILE ASSEMBLING THE hand sleds, Spencer saw the motorsleds bearing Demorest and Tetley approaching from the east, each with a cargo sled in tow. They followed the path of ski tracks they'd made the night before, with Demorest out front and Tetley close behind.

Motorsleds are heavy and difficult to turn. Demorest wanted his to be facing in the opposite direction, away from the B-17, when the time came to leave. About one hundred yards from the bomber, Demorest turned off the trail and steered the motorsled in a wide arc. When he reached the end of the arc, Demorest expected, he'd be pointing away from the PN9E. Even after Spencer's fall into the crevasse, Monteverde's crew had walked around the crash area to gather airdropped supplies. Demorest must have thought the well-trod area was safe from danger.

But twenty-five yards into the turn, Demorest's motorsled crossed a snow bridge over a hidden crevasse. The bridge might have supported a man alone on foot, but it was too weak to bear the weight of a man on a loaded motorsled. Without warning, the bridge gave way. Max Demorest and his motorsled plunged head-first into the blue-white abyss, with the attached cargo sled pulled in behind.

Tetley jumped off his sled and sprinted to the crevasse, yell-ing for help as he ran. When he reached the eight-foot-wide open-ing, he looked down to see the motorsled's tail and the cargo sled, wedged between the walls about one hundred feet down. But he saw no sign of Demorest. He shouted for his friend but heard no answer.

Grabbing ropes, Monteverde, Spencer, and Howarth ran to join Tetley at the hole. They called to Demorest, but received no reply. Howarth volunteered to trek as quickly as possible to the Duck's landing site to tell Pritchard and Bottoms what had happened. He'd ask the Coast Guard pilot to return to the *Northland* to gather more men and equipment to attempt a rescue. Monteverde gave him the go-ahead and told Howarth that Pritchard should take off as soon as possible, without waiting for more crewmen from the PN9E. They needed help fast, and snow and fog were rolling in from Koge Bay.

Howarth stopped at the tail section for a fast good-bye to Bill O'Hara, Paul Spina, Clint Best, and Clarence Wedel. They wished him luck, told him to avoid the crevasses, and gave him eight rolls of film they'd taken of themselves. They told Howarth they wanted him to have sets of photos developed for each of them aboard the *Northland* when they arrived. Howarth tucked away the film and followed the trail Pritchard and Bottoms had made the day before to the Duck.

At the crevasse, Tetley frantically wove a rope ladder. Mon-teverde considered lowering a man, but this was much different from Spencer's fall. With no sign of life from Demorest, Monte-verde concluded that sending a man down would recklessly en-danger another life. Tetley and Spencer agreed. Trying to reach Demorest would take every healthy man among them, and even then they might not have enough strength or equipment. They had no option but to keep watch and continue calling to the lieu-tenant. If he answered, they'd send a man down.

One PN9E crewman rushed to the radio to call for help, but without Howarth's knowhow, all he could do was send SOS messages. The *Northland* heard them but misunderstood their purpose. The ship's radio operator relayed a message to "Iceberg" Smith saying that the bomber was "apparently attempting to contact motor sledges in her vicinity or [at] Ice Cap Station."

JOHN PRITCHARD AND Ben Bottoms landed the Duck at the same spot where they'd come down the previous day, some two miles from the PN9E. This time, Pritchard landed on the Duck's pontoon, with the wheels up, as though the ice were water. Bottoms radioed the *Northland* that they were safely on the glacier.

When Howarth reached the landing area, he told the Coast Guardsmen about Demorest's fall. Pritchard and Bottoms brought the young radioman aboard for the return trip to the *Northland* and prepared for an immediate departure. Pritchard and Bottoms climbed into the cockpit and Howarth crawled down into the empty compartment in the fuselage.

Pritchard took off in the same direction as he'd flown away the day before, heading toward Koge Bay. As he flew south and passed over the wreck, Pritchard waggled the Duck's wings in salute.

Below, at the crevasse, Monteverde, Spencer, and Tetley were still calling to Demorest. He never answered. Soon the fog grew so thick that they had to abandon their vigil, lest they risk falling into a crevasse, too. They returned several times to the hole when the fog cleared, but heard nothing but the echo of their own voices.

Two years earlier, Max Demorest's mentor had published a worried insight about his protégé's fearlessness. Now the prophecy had come true for Max Harrison Demorest of Flint, Michigan, a thirty-two-year-old Ivy League professor and scholar, husband and father, army lieutenant and motorsled rescuer.

While trying to help a group of desperate men, one of the

world's leading authorities on glaciers had been lost forever at the bottom of one.

WITHIN MINUTES OF takeoff, the Duck and the three men aboard were in trouble, too. The storm had arrived faster than Pritchard expected, and the window for a safe return to the waiting *Northland* in Comanche Bay had closed.

Roughly nine minutes into the flight, when the Duck should have been about halfway back to the ship, Pritchard called the *Northland* for a weather report. From the little radio room a deck below the bridge, the ship's communications officer began to reply. But before he could finish, Pritchard cut him off, yelling, "MOs, MOs—quick!"

The urgent call could mean only one thing: Pritchard was lost, disoriented in the fog and storm, flying at perhaps ninety knots, or about one hundred miles per hour. Pritchard was flying in milk. By calling for magnetic orientation, he was desperately seeking a course to the *Northland*. Without it, the Duck was in danger of slamming into the water or the ice cap. Pritchard was the airborne equivalent of a sailor searching for a beacon to guide him past a reef.

The *Northland*'s radio operator hammered his transmitter key, sending the signal for MOs—five dashes in succession, da-da-da-da-da—on a prearranged frequency. He repeated the signal again and again—da-da-da-da-da, da-da-da-da-da—but received no reply. The *Northland* was trying to tell Pritchard and Bottoms that the course was 115 degrees true, a heading over the ice-covered landscape to the waiting ship. But the radio calls to the Duck remained unanswered, and the little plane never emerged from the fog.

The crew of the *Northland* knew that Pritchard, Bottoms, and Howarth were in danger, or worse, but they refused to write them off. Maybe Pritchard had turned back toward the PN9E to land on the ice cap. Maybe, *Northland* crew members told themselves, he'd landed somewhere else safely but his radio had been damaged. Or maybe the Duck had set down on the water and was floating in Koge Bay. Perhaps the Duck's radio signal was blocked by the storm, and that's why no one could reach them.

AFTER STRUGGLING TO use Howarth's patched-together radio, Don Tetley told the *Northland* the bad news at the crash site: "Demorest and one motor sledge in crevasse. Unable to get help to him. Please send help immediately. . . . No word from *Northland* plane."

In reply, the ship told Tetley that the Duck hadn't returned to the *Northland*, either.

News of the plane's disappearance fell hard on the PN9E crew, most of all O'Hara and Spina. That morning, before the Duck left with Howarth, the other crewmen had helped to dress the two most seriously injured men, expecting that they'd be next to leave. With the Duck nowhere to be found, and with one motorsled gone, O'Hara and Spina seemed to be running out of ways to leave the glacier for medical care. Maybe running out of time altogether.

Still, the men on the ice and on the *Northland* retained hope that Pritchard, Bottoms, and Howarth were alive, and that they'd landed somewhere between the PN9E and the ship. Until there was proof one way or another, the Duck's crew and the PN9E's radioman were missing, not dead.

During the next few days, Greenland Patrol commander "Iceberg" Smith exchanged a series of messages with the *Northland* in which he expressed concern for the lost Duck and told the ship to leave Comanche Bay "if by remaining, ship and personnel are endangered." The PN9E rescue was supposed to take about three days, and it had already stretched longer. Each day the *Northland* waited, the threat of ice capturing the ship in Comanche Bay increased. A follow-up message from Smith delivered the point more sharply: "Emphasize that remaining to continue rescue operations must be secondary to the safety of *Northland*." The ship was his primary workhorse along the coast, and he couldn't afford to lose it. And yet, despite his doubts, nowhere in the radio messages did Smith question the decisions by the ship's top officers or by pilot John Pritchard.

A WEEK AFTER the Duck's last flight, an American B-17 flew over the Koge Bay glacier. Its pilot, Captain Kenneth Turner, reported a sad but unsurprising discovery: "Grumman [Duck] located. No sign of life. Badly wrecked."

The broken remains of the plane were about three miles from the water, Turner said, although later sightings claimed that he overestimated the actual distance. No one would question Turner's heartbreaking description of what he saw. The Duck's tail pointed skyward and its wings were broken off. Its fuselage was intact, but "front part of ship demolished."

The evidence pointed to a nose-first dive. Still, investigators, Coast Guard officials, and armchair historians would speculate for decades about what happened. Above all, they'd argue whether Pritchard had tried to turn back to his takeoff area when he ran into the fog, or if he'd continued heading for home, toward the *Northland*. The latter seemed more likely to most, as the navigator for Turner's B-17, Herbert Kurz, created a map that showed the downed Duck pointing in the direction of Comanche Bay.

However it happened, the sighting by Turner and his B-17 crew established that Pritchard, Bottoms, and Howarth were gone and the Duck was destroyed.

In the end, Tetley's initial radio message—"Demorest and one motor sledge in crevasse. . . . No word from *Northland* plane"—disclosed the failure of two rescue methods, by air and by motorsled. More important, it reported the deaths of four American heroes, all of whom had died after volunteering to help fellow servicemen.

In the weeks that followed, a Coast Guard team from the *Northland* attempted to reach the little plane. Yet from the very first, there was no hope that the three men had survived. Still, one overly optimistic message from the *Northland* five days after the crash requested that the Duck be dropped a cache of supplies and tools, including a new battery, sixty gallons of fuel, eighteen spark plugs, and several wrenches that would be needed to prepare the engine for takeoff. The message also requested three sleeping bags, animal hides, coffee, food, and spirits.

In reality, after Turner spotted the Duck, all the efforts to reach the plane were motivated by a desire to recover the three men's remains, to honor their sacrifices, and to grant peace to their families. But storms, billowing snow, and uncertainty about precisely where to look forced the searchers to abandon the effort. Pritchard, Bottoms, and Howarth were left where they fell.

In March 1943 Colonel Bernt Balchen also flew over the wrecked Duck, afterward drawing a rough map of its location, with an X marking the plane's spot. The result was a sketch that looked like a pirate's treasure map. Although the map was accurate in several important respects, Balchen invented several geographical features and omitted others. Complicating any future search for the Duck and its men, Balchen's errors compounded flawed longitude and latitude data reported by Turner's B-17 crew. Some of those errors were later corrected, but the combined effect was to create confusion and false leads about the Duck's resting place. That problem would reverberate for decades, as a steady accumulation of snow and ice buried Pritchard, Bottoms, Howarth, and the Duck.

AFTER THREE MILITARY plane crashes in November 1942 in which all men aboard initially survived—McDowell's C-53, Monteverde's PN9E, and the Canadian A-20—Greenland had struck back. In less than two hours on the morning of November 29, 1942, a Coast Guard pilot, a Coast Guard radioman, an Army Air Corps radioman, and an army lieutenant motorsled driver were killed trying to save others.

That death toll soon rose, as the search for McDowell's C-53 crew was abandoned a month after their crash. "Concentrated search was discontinued," the official report declared, "because it was believed that the crew could not maintain life more than thirty days with the short rations they had on board, and during which time extremely cold weather prevailed. It is believed the

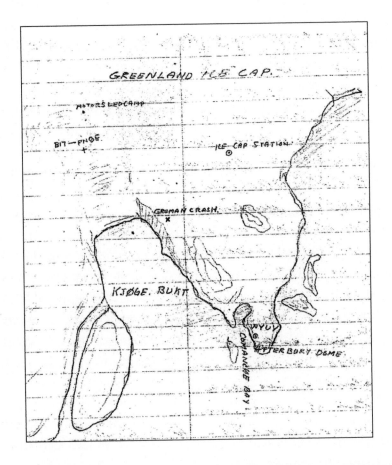

THE MAP COLONEL BERNT BALCHEN DREW IN HIS NOTEBOOK AFTER FLYING OVER THE WRECKED DUCK AND THE KOGE BAY AREA. *(U.S. COAST GUARD IMAGE.)*

crew perished. Aircraft is considered a total loss due to the inaccessibility of the Ice Cap, should it ever be located." Official declaration of their deaths would come on November 5, 1943.

Frozen in time in November 1942 were Lieutenant John A. Pritchard Jr., Sergeant Benjamin Bottoms, and Corporal Loren "Lolly" Howarth of the Duck; Captain Homer McDowell, Lieutenant William Springer, Staff Sergeant Eugene Manahan, Corporal William Everett, and Private Thurman Johannessen of the C-53; and motorsled rescuer Lieutenant Max Demorest.

As Greenland's god-awful winter approached, seven men remained trapped on the ice: Armand Monteverde, Harry Spencer, William "Bill" O'Hara, Alfred "Clint" Best, Paul Spina, and Clarence Wedel from the PN9E, and their new companion, motorsled driver Don Tetley.

In a battle against nature, fought at the far edge of a war among men, Greenland had regained the upper hand.

POSTHUMOUSLY, JOHN PRITCHARD and Benjamin Bottoms each received the Distinguished Flying Cross. Both also were recommended for the Medal of Honor, the nation's highest military decoration, but the award was never made.

Pritchard's Distinguished Flying Cross citation honored him "for heroism and extraordinary achievement while participating in aerial flights as pilot of a plane which rescued Army fliers stranded on the Greenland Ice Cap, on 28 and 29 November 1942. . . . By his courage, skill and fearless devotion to duty, Lieutenant Pritchard upheld the highest traditions of the United States Naval Service." The medal was presented to his parents in a ceremony at their Congregational church in Los Angeles, with music by Rudy Vallee and the Coast Guard Band. For the loss of her firstborn child in heroic circumstances, Virginia Pritchard was honored as the "California Mother of the Year" for 1944.

Ben Bottoms's Distinguished Flying Cross citation read, "He

rendered valuable assistance to the pilot on the two flights to the Ice Cap, maintained excellent contact by radio between his plane and their ship, and assisted the pilot in rendering aid to the injured and stranded fliers."

For repairing the radio at the risk of his own life, and for trying to help Max Demorest when he fell into the crevasse, Loren Howarth received the Legion of Merit, the sixth-highest military award, for "exceptionally meritorious conduct in the performance of outstanding services and achievements."

WORD OF THE Duck's crash reached Al Tucciarone and Woody Puryear while they were recovering aboard the *Northland*. Several weeks after his rescue, Tucciarone gave military investigators a sworn account of the PN9E crash and its aftermath. At the end, he veered from formal chronology and wrote, "I want to stress that I owe my life to Howarth, Pritchard, and his radioman, Bottoms." Thinking about the men still awaiting rescue on the ice, Tucciarone wrote, "I can only hope and pray that those other unfortunate fellows get away from that 'Death Hole' as lucky as Sergeant Puryear and myself. May God bless them."

From his hospital bed, where he'd remain for more than two months, Puryear wrote to Pritchard's and Bottoms's families to offer thanks and condolences.

To Bottoms's parents, he wrote, "I am one of the boys whose life was saved through the heroic efforts of your son, Benjamin A. Bottoms, and Lieutenant John A. Pritchard. Two braver men I have never seen. I knew your son for only a short while and had never seen him until the day of the rescue. He was more than willing to go to the limit to save our lives, even though endangering his own."

After receiving a similar letter from Puryear, Virginia Pritchard wrote back, "I breathed a little prayer of thankfulness when I learned you were back in this country and able to wire. We know how very ill you were from hunger and exposure. We have no hope whatever that our son still lives, but until the final chapter is written, we have faith that somehow a miracle will bring him back."

Virginia Pritchard died in 1976 with that prayer for a miracle unfulfilled.

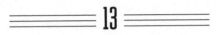

13

TAPS

WALKING THE CORRIDORS of Coast Guard Headquarters in Washington, Lou looks more excited than nervous, a prizefighter heading toward the ring. Alongside him is John Long, who retired from the Coast Guard in 2011 but remains deeply involved in Duck Hunt research. Long is an ideal cornerman: steady, reserved, a font of knowledge about Duck sightings. The size of a longshoreman, the former master chief petty officer flew to Washington from his home in northern Michigan to lend his support and expertise. They walk shoulder to shoulder into the boardroom.

The seats around the table are filled with many of the same men and women from the October meeting at the DPMO offices, along with a half-dozen new faces from the Coast Guard and other agencies. Leading the DPMO delegation again is Lieutenant Colonel James McDonough, a friendly but no-nonsense soldier.

Coast Guard Commander Jim Blow kicks things off with a positive spin: "It's really good to get all the players in one room to see what we as a group can do to pull the pieces together to get on site and effect a location and a recovery."

Blow hands off to Lou, who's loaded for duck this time. He distributes a nineteen-page memo that's thick with data suggesting that he knows where on the ice to dig for the plane, along with day-by-day plans for the mission. Lou guides the room through a PowerPoint presentation, confidently fielding questions and explaining a thicket of technical information. He describes primary and secondary locations where modern radar and sonar findings overlap with historical sightings, including Bernt Balchen's detailed, if geographically imprecise, hand-drawn "Treasure Map" from 1943.

McDonough wonders how much of the Koge Bay glacier Lou intends to search. "What dimensions are we talking about?" he asks.

Lou: "One and three-quarters of a mile by three-quarters of a mile, tops."

Next McDonough jabs Lou about whether anomalies labeled "Points of Interest" on a radar readout are definitely metal objects. Or, he wonders, are they natural variations within a glacier made by a rock, a crevasse, or even water?

Lou counters: "Metal comes back differently than rock."

Lou explains that, despite his confidence in the available data, he hopes to confirm the findings with additional high-tech devices, one that hangs beneath a helicopter and another that drags behind a snowmobile. "We have the technology to find it," Lou says. "We have a very limited area in which to look. We will find it."

McDonough asks: "So, you think you can find the aircraft in three days on the ice?"

Lou's on a roll. He snaps back: "Weather permitting, yes."

The give-and-take moves to what Lou might do with the plane after carving it from the ice. "We're not wreck hunters," he says. This is a point of pride with him, and he puffs his chest a bit. "We're not interested in recovering the plane to sell it."

Lou says he respects the Coast Guard's policy that the service retains possession of all downed planes and shipwrecks. His only desire would be for the Duck to be restored and displayed for posterity. This, of course, is exactly what retired captain Tom King wanted in the first place, to keep it off eBay and away from private warbird collectors and souvenir vultures.

McDonough says DPMO would have no objection to that plan. "The plane is not the priority," he says. "Our primary interest is in the remains. What I'm wondering is, will the bodies be encased in ice?"

Lou: "If snow has gotten into the Duck, yes, they're probably encased in ice."

McDonough: "Could you bring it up without disturbing the remains?"

Lou: "That's possible."

He explains how *Glacier Girl* was brought to the surface twenty years earlier, and tells McDonough that this would be easier because the Duck should be located at less than one-fifth the depth. A friendly discussion follows about the best ways to recover, preserve, and transport human remains. The difference in tone from the October meeting is stark.

I start to believe that Lou is winning. I imagine McDonough raising Lou's arm in triumph and presenting him with an oversize check while cameras flash. But then the bell rings for the next round.

McDonough asks: "Lou, do you have an underwriter at the moment?"

For the first time, Lou pauses to catch his breath. "No," he says flatly. "That's what we're working on." What he means is, That's why we're here.

McDonough bores into the budget that Lou included in his briefing package, asking about how large a team he would bring, whether he'd have a medical officer for emergencies, how he'd get supplies to the campsite, even whether he'd feed his team with military meals-ready-to-eat, MREs, or commercially purchased freeze-dried foods. Lou tries not to show it, but the meeting has gone more than two hours, and the unrelenting questions have a desiccating effect on him. His answers come more slowly.

McDonough circles back to the question of money and drops the hammer. "To be honest with you," he says, "a lot of us were under the assumption that you had found private funding to get out onto the ice."

Lou tries to regroup. He explains that he's been seeking support from corporate and media sponsors and private individuals, though none are firmly on board. This is the source of McDonough's misunderstanding: During the weeks leading up to the meeting, Lou sent group e-mails with glowing reports of sponsorship discussions he's had with wealthy World War II history buffs and others. McDonough interpreted the e-mails to mean that Lou had landed a big fish.

Jim Blow steps in, trying to help Lou: "When do you think you'll have something back from the private funders?"

Lou is reeling. He can't see that Blow is trying to help, and he answers flippantly: "That's the sixty-four-thousand-dollar question, I guess."

McDonough is ready to end it with a knockout. He wants no further misunderstandings or, apparently, further meetings: "I think it's very admirable what you've done," he says. "I want to compliment you. But funding is always the long pole in the tent. . . . I don't believe there's any funding available on our end, certainly not this fiscal year, to support an investigation on the ice.

"In a perfect world, we'd go after everyone we can get," McDonough adds. "But the reality is, the money isn't there."

As an aside, he offers a small ray of hope: if Lou can somehow get to Greenland, and if he can confirm beyond any doubt that he's found the Duck and its occupants, the Defense Department would be obliged to get involved.

McDonough's comment is a sideways reference to a congressional mandate. In 2010, under pressure from families of missing servicemen, Congress told the Defense Department to speed the pace
of MIA recoveries. Specifically, federal lawmakers amended a law
known as the Missing Persons Act, or Title 10. In the amendment,
Congress ordered the creation of a "comprehensive, coordinated, in-
tegrated, and fully resourced program to account for designated persons who are unaccounted from World War II, the Vietnam War,
the Cold War, the Korean War, and the Persian Gulf War." Congress
also required that by fiscal year 2015 the Defense Department spend
enough money to bring home at least two hundred MIAs annually,
a sharp increase from the current yearly average of eighty-five.

Later, McDonough reiterates the point in an e-mail: "Don't forget what I said about Title 10. Should you get onto the ice and
make a discovery it would be a game changer."

But as the meeting draws to a close, McDonough focuses on
delivering the bad news, knowing that Lou sees silver linings in
the darkest clouds. He defines the current situation: "It's January
now. You're looking at going there in May. If you don't have an
underwriter, you won't get up there?"

Lou acknowledges: "It's looking that way."

Yet Lou still won't surrender, offering one more shot that falls
somewhere between a pitch and a plea: "Where do we go from
here? Can anybody go back and look at this? Why stop now?" He
tries to build momentum: "There's a lot of congressional interest
in this. Have you seen the letters they've sent?"

This smacks of desperation. Everyone at the table knows that
the real trick would be finding a member of Congress who'd publicly *oppose* retrieving World War II heroes. Joan Baker, a forensic
anthropologist for DPMO, rolls her eyes when Lou mentions the
letters. Wearing the expression of a person who thinks her time is
being wasted, Baker answers coolly: "We see a lot of letters in our
office. Perhaps some of the congressmen can contribute."

Lou says softly, "I can find these guys." Speaking more to himself than to the dwindling crowd, he adds: "It's just a question of money."

FEBRUARY–MARCH 2012

The phone rings four times before a woman answers.

"Hello?"

"Hi, Nancy, it's Mitch. Calling again about John."

I quickly get to the point: "Nancy, I have a question. If John's body were found, where would you want him buried?"

"Oh. Let me think a moment," Nancy Pritchard Morgan Krause says in her lilting voice. "He was Coast Guard, so it would be nice if he were returned to the Coast Guard Academy. But they have to find him first."

I update her on the financial and other hurdles facing Lou, North South Polar, and the Duck Hunt. I thank her and say good-bye.

I met Nancy seven months earlier, at her retirement home in Annapolis, Maryland. At eighty-eight, she's a trim, lovely woman with snow-white hair that she keeps short and stylish. Nancy and her second husband, Bill Krause, who's ninety, are competitive croquet players who enjoy travel, easy banter, a civilized cocktail hour, and each other's company.

Nancy married Bill after the death of "Tick" Morgan, her brother John's best friend. Tick died in 2004, shortly before he and Nancy planned to celebrate their sixtieth wedding anniversary with a big family reunion. She turned the reunion into a memorial. "That was closure for my husband," Nancy says. "We're still waiting for that with my brother John."

The day we met, as Bill served drinks, Nancy ran her long, elegant fingers over the cover of the Coast Guard Academy's 1938 yearbook. Its spine is like a hinge to the page with "Johnny"

Pritchard's entry: two years of football, two years of boxing, year-book staff, newspaper staff, basketball manager. Then a short pro-file of her eldest brother that Nancy has read too many times to count: "Before you is a product of the fair state of California—a person whose disposition bears out the reputation of that state for sunshine. His ready smile and overflowing chatter have not only become a tradition at the Academy, but have buffaloed many a member of the fair sex into believing all his promises."

Nancy smiles as she talks about her "confident, self-assured" big brother, nine years her elder, and about how gentle and caring he was toward her. Nearly seventy years after the fact, she cries when she describes the phone call she received from her mother while at college. "She said, 'Nancy, John's been lost.' That was it." Nancy left her dormitory, went out into the falling snow, and walked around the block, knowing that she'd never fully recover from the loss.

With her parents and other brothers gone, Nancy is John Pritchard's closest surviving relative, what the U.S. military calls his PNOK (pronounced "pee-knock") or primary next of kin. That gives her final say over where his remains would rest, should they be recovered. In 1975, when the Coast Guard first tried to find the Duck, Nancy was skeptical. "I said at the time, 'Leave him there. Let him rest in peace. That's where he went down, and I would hate to see anybody else put in danger.'"

Now Nancy feels differently. "Congress has said they want all the MIAs, the missing in action, to be brought back to this country, and I agree. If they bring everybody back, then by God, you bring my brother back."

AT AGE SEVENTY-SEVEN, Edward "Bud" Richardson still sees his stepfather Benjamin Bottoms through the eyes of a small boy.

"The biggest thing I remember about him is that he taught me not to be prejudiced," Bud says. "Sometimes he'd show up at home with two or three soldiers or sailors, whoever was at the bus stop that night, looking to go into town to have a few drinks or whatever. He'd bring them home for dinner instead.

"I recall him bringing home a black man, and I had never seen anybody other than people who were white. I must've looked surprised, and he told me, 'There are different people in this world, different colors, different eye shapes, but they're all people. That's just how God made them.' That's the biggest lesson I learned from him, and I remembered it always."

The rest of what Bud knows about Ben Bottoms is from snippets and snapshots, some drawn from his own memory, some from stories told by his mother. Bud remembers Ben teaching him to swim in the ocean off Gloucester, Massachusetts; riding on his shoulders to buy an ice cream cone; losing his sailor's cap when they ran home to avoid a storm; unwrapping skis from Santa Claus. Bud always knew that the skis came from the other man in his life with a bushy beard.

Though unrelated by blood, these days he bears a distinct resemblance to his stepfather, with a rounded face and a receding hairline. He remembers his mother, Olga, a pretty woman with coal black hair, refusing to accept that her husband was dead. "She had a belief that maybe he was alive up there, maybe Eskimos rescued him." Bud says she only relented after a Coast Guard officer assured her that a pilot had seen the wrecked Duck and the bodies of its crew. The part about the bodies was doubtful, but it had the desired effect, putting to rest Olga Bottoms's dream that Ben had survived.

"As a young boy, I had illusions about going up there to Greenland and bringing his body back," says Bud, a retired construction manager and stable owner. As he got older, he considered going to the Coast Guard Academy, but that plan washed away when his mother married a navy officer. Bud thinks the marriage was mostly designed to give him a secure home and a father figure.

If his stepfather were found, Bud says, he'd probably be buried at Arlington National Cemetery. But Bud wonders if instead Ben Bottoms should rest in his native Georgia. Either way, Bud wants him home. "I feel very good they're going back to get him," he says. "It's just a shame it couldn't have happened when my mother was alive. He was the love of her life."

One more memory: when Bud was a boy he learned to play the trumpet, but the one song he could never play was "Taps."

"I played it once and she totally broke up. I never did play it again. I knew why."

JERRY HOWARTH WAS born in the same four-room log cabin in Wausaukee, Wisconsin, as his uncle Loren "Lolly" Howarth. Jerry was not yet two years old when the Duck crashed, so all he knew of his uncle were stories from his father, Loren's younger brother.

"Loren was basically a country boy," Jerry says. "Everybody worked together on the family farm. Lived off the land. There wasn't much money to make. They'd hunt and fish. Deer, mostly. Duck, too.

"Everybody said he was awful quiet."

There's pride in his voice when Jerry says Loren was the first member of the family to go to college. "He washed dishes and worked in restaurants to make his way."

As Loren Howarth's PNOK, Jerry provided the Coast Guard's John Long with a wristwatch that belonged to one of Loren Howarth's brothers, from which a DNA sample was taken. "I wish my dad and his brothers and all of them were still here, but it's a good idea to bring him home where he belongs."

Marc Storch, a cousin by marriage, is the family historian and the keeper of Loren Howarth's Legion of Merit, which he inherited from Loren's widow, Irene.

"When she first showed it to me, she said it was Lolly's and she smiled," he says. "She said, 'He was such a sweet boy.' Remember, Irene was talking about someone who stopped growing old in 1942. There she is at one hundred and one years old. Lolly never got any further than his twenties. So Irene, eighty years later, is still seeing that boy, that young man.

"It means so much that Lolly could come home and be close to where his family is," Marc says. "Even though it's only a physical reuniting, having his remains back here would be important to those who remember him. It would also be important to those who know what he did to help save his crew.

"People should know about that, and what it cost him."

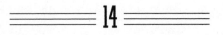

14

GLACIER WORMS

DECEMBER 1942

As November turned to December 1942, the days grew shorter, the nights colder, the survivors' hopes dimmer. The six icebound members of the PN9E crew, now joined by Don Tetley, faced the awful truth that their two best chances for rescue had gone down with Max Demorest's motorsled and John Pritchard's Duck.

Their spirits fell even lower during the first week of the new month. Heavy storms with windblown snow made it almost impossible to leave the bomber's tail. Rations ran low as no supply planes could reach them.

In addition to Tetley, the remaining men trapped on the ice were pilot Armand Monteverde; copilot Harry Spencer; navigator William "Bill" O'Hara; engineer Paul Spina; passenger Clarence Wedel; and volunteer searcher Alfred "Clint" Best.

Time and hardship had revealed Monteverde to be confident enough to take advice freely and to give orders only when necessary. Spina considered "Lieutenant Monty" to be a hero for the way he held them together.

Although Spencer was the youngest crew member, he had the traits and the touch of a natural leader. Even after falling into the crevasse, he was the strongest and most capable among them, a likable fellow with sensitive radar for when one of his crewmates needed an extra ration or a supportive shoulder.

To a man, they admired O'Hara for his tough-guy stoicism, even as his numb, discolored feet worsened and the blackness spread up his legs.

They valued Spina for his relentless good cheer despite his injuries and agonizing frostbite. Even when Spina moaned about pain in his hands and feet, he did so with the timing of a vaudeville comic. Spina's comfort in tight quarters might be traced to the fact that he was the third of seven children of a homemaker and an Italian immigrant factory worker.

Wedel, a stranger to the others just weeks earlier, had earned respect for his mechanical ingenuity, somehow fixing their tem-

PRIVATE CLARENCE WEDEL. *(COURTESY OF REBA GREATHEAD.)*

peramental generator despite frozen parts. Powerfully built, with
dark, wavy hair, a cleft chin, and bright blue eyes, the thirty-five-
year-old Wedel was one of the more unusual privates in the U.S.
Army.

Born on a Kansas farm, the eldest of ten children, he was raised
a Dunkard, a tiny Christian denomination of pacifists whose
members, like Mennonites and Quakers, could claim exemption
from military service. But Wedel believed that it was wrong to
use his religion to avoid the war. Six months after Pearl Harbor,
Wedel left the welding business he owned with his father and
enlisted. He left behind his pregnant wife, Helen, a violinist ten
years his junior whom he'd married on Christmas Day 1941. The
two shared a love of dancing, and they'd spent their honeymoon in
the "big city" of Wichita, at a nightclub named after their favorite
song, "Blue Moon."

Clint Best was easygoing and introverted. He had no bluster
or bravado, and he won praise for mixing the crew's monotonous

TECH SERGEANT ALFRED "CLINT" BEST. *(COURTESY OF ROBERT BEST.)*

rations into creative meals. But Best was no outdoorsman, and he was perhaps the least suited among them for the deprivations of Arctic survival. The son of a traveling shoe salesman turned grocer, Best was happiest working inside with numbers. Equipped with a layer of padding from years at a desk, the brown-haired, blue-eyed Best had worked as a bookkeeper for a wholesale distribution company in Memphis before the war. During the five months he'd been in Greenland, cracking codes in a heated office at Bluie West One had been a perfect fit. Being cold, hungry, and trapped in an oversize icebox, watching men disappear into crevasses and going down in airplanes, was torment for the cryptographer turned volunteer searcher. As days passed, Best retreated into his own thoughts.

The newest member of their band was Tetley, a wiry Texan who fit the stereotype of the quiet cowboy. After Demorest's

SERGEANT DON TETLEY. *(U.S. COAST GUARD PHOTOGRAPH.)*

death, Tetley drove his motorsled over the crevasse-free ski tracks
and parked alongside the wrecked PN9E. He'd been trained by
Demorest in Arctic life, and even a short time in the cramped tail
section made him seek alternative lodgings. It wasn't the crowding—he was used to that from living at Beach Head Station and
Ice Cap Station—it was the precarious position of the fuselage.
Although secured with ropes to the front half of the plane, the tail
perched over an expanding crevasse similar to the one that killed
Demorest. On Tetley's first night in the fuselage, he was startled
when the tail section shifted. Fearful of sliding into the abyss, he
climbed out of his sleeping bag and declared: "I'm going out and
dig myself a hole [to] crawl in."

WITH MORE STRENGTH and energy than the others, Tetley dedicated himself to the tasks of improving their lodgings and plotting a way out. He converted the metal cover of the PN9E's
Norden Bombsight into a crude saw and carved out blocks of
snow under the bomber's unbroken right wing. Spencer and Wedel pitched in, using a jungle knife, a shovel from Tetley's supplies, and tools from their mess kits.

Within several days they'd dug a "room" with walls of ice
about fifteen feet long, eight feet wide, and more than four feet
high. The roof was the metal underside of the wing. They couldn't
stand straight, but at least they could stop living like sardines and
crevasse-bait. Upon moving from the tail section to the underwing
snow cave, the seven men spread out sleeping bags to their full
length. One drawback was that the ice underneath them melted
from their body heat, soaking the sleeping bags with no way to
dry them.

When they'd all moved in, Tetley set up his stove in the cave.
The men held their breath when he lit it, fearing the fuel-filled
wing above them. The metal pinged and moaned when it first
heated, but it posed no danger. Wedel made the cave homier by

stringing a lightbulb on a wire from his generator. The well-lit, white-walled room brightened their spirits.

With Howarth gone, Tetley became the new radioman, with Best as his assistant. They lacked Howarth's communications knowledge, so they couldn't get the transmitter to achieve its full range. They could send messages only by Morse code, but they could receive incoming voice transmissions. Despite Wedel's unceasing efforts, the generator was unreliable, so the radio and the light were on-and-off pleasures.

The men blamed mechanical woes, missing items, and other unexplained troubles on "Glacier Worms." There were, in fact, creatures called ice worms that lived in glaciers, though not in Greenland. But in the stranded men's imaginations, Glacier Worms became the ice cap equivalent of gremlins: mischievous, mythical beasts that bedeviled airplanes in flight and, now, on the ice.

With their new quarters complete, the PN9E survivors cut the lines securing the bomber's tail section. Their home for the previous four weeks slid into the crevasse with a thunderous roar and disappeared from sight.

AFTER SEVERAL DAYS of hoping the Duck might return, the men of the PN9E cast aside lingering dreams of being airlifted to safety. Winter was closing in and the cavalry wasn't on the way. They reported a temperature of 16 degrees Fahrenheit and sent requests via the *Northland* for supply drops: "We need food. . . . Everyone OK, but weak." They made other requests, as well: "If [supply] plane comes . . . we need flashlight batteries, laxatives, bandages, candles, and reading material."

The supplies arrived in an airdrop from the B-17 flown by Captain Kenneth Turner. A Salt Lake City native approaching his fortieth birthday, Turner was mature, balding, and composed. He seemed ancient to the young flight crews he worked with, so everyone called him "Pappy." Like the PN9E, Turner's B-17 was in the temporary possession of the Air Transport Command on its way to England. Also like the PN9E, it had been diverted from its destiny as a weapon into the role of a search-and-supply lifesaver.

Supply drops by Turner and his crew satisfied the immediate needs of the men on the ice, but they couldn't stop O'Hara from getting worse by the day. As Monteverde changed the dressings on the navigator's feet, he grew convinced that little chance remained of saving them. O'Hara also was losing more weight than the rest of them. He could stomach only a few drops of thin soup. Spina needed expert medical care, as well, and the others worried that neither man might last long.

Despite the commitment by Pappy Turner and his crew to drop supplies whenever possible, the men in the snow cave feared that the approach of winter might block resupply efforts for weeks. While O'Hara was asleep, Tetley told the others that they were gambling with the navigator's life if they thought they could spend the winter relying on supply drops and waiting for a rescue party. Monteverde agreed, so Tetley radioed the *Northland* with the first draft of a plan to take matters into their own hands.

"In case of emergency, we could travel light," he tapped out in code. "We believe that our seven-man party could reach Ice Cap Station on our one motorsled. Would travel slightly altered course to avoid crevasses in this area. Could meet dogsled on trail."

Initially, the ship instructed Tetley to sit tight and await another dogsled team heading their way. In the meantime, the *Northland* would send ashore its hospital corpsman to provide medical aid if the PN9E crew could reach Ice Cap Station or Beach Head Station with help from the dogsled team. But that plan soon changed. Again the dogsled turned back, unable to make it through the driving, drifting snow. So the men under the wing of the PN9E plotted to save themselves.

On December 7, the one-year anniversary of the United States' entrance into the war, the weather broke. The sun shone and the wind died down. Such a rare fine day in Greenland might not appear again until spring. Monteverde concluded that O'Hara could wait no longer. He reasoned that hauling all seven of them to Ice Cap Station on the one remaining motorsled would be difficult at best, suicide at worst. But if they split up, maybe two smaller crews could survive separate trips.

At first, Monteverde wanted both O'Hara and Spina on the first outbound motorsled. But after discussions with Spencer, Tetley, and the others, he abandoned the idea of sending the two most seriously injured men onto the ice cap at the same time. Monteverde decided instead that the first group would be a four-man team: O'Hara, wrapped in a sleeping bag, would be strapped on the supply sled towed by the motorsled; Tetley would drive because he knew the machine and the route; Wedel would provide strength and mechanical skills if the sled broke down; and Spencer would be in command, doubling as the navigator if they got lost. With three able-bodied men and one badly injured man, Monteverde reasoned, they'd have enough muscle to push or tow the motorsled out of a snowbank. As soon as the weather allowed, Tetley would return on the motorsled for the three remaining men: Monteverde, Spina, and Best.

With an escape plan in place, they gathered for a group prayer. They offered the travelers good tidings; wished the men remaining behind a short stay; and whispered blessings for Demorest, Pritchard, Bottoms, and Howarth.

Tetley radioed their revised plan to the *Northland*: "Lieutenant O'Hara very ill. Leaving with him . . . within an hour." He didn't wait for approval, and he never formally received it. But he did get tacit support when the *Northland* instructed the men at Ice Cap Station to turn on their lights as a beacon.

Having already traveled the thirty miles between the PN9E and Ice Cap Station, Tetley thought they could make the trip in a single day. This time, though, he'd follow a route recommended by Colonel Balchen, who'd mapped it out from the air. Balchen's course steered them to the north, away from the crevasses. In case the trip took longer than he anticipated, Tetley gathered three days' rations, sleeping bags, a shovel, and a tent. That left the men in the snow cave equipped with Tetley's stove, fuel, a second shovel, his walkie-talkie, and other supplies if his return was delayed. One piece of bad news was that Wedel couldn't start the generator, so he couldn't leave the cave dwellers with three fully charged batteries for the radio, as he'd hoped.

The thought of taking action energized them all. Before they parted, they joked around and wished each other well. Tetley and Spencer said they'd be back within two days, to celebrate Monteverde's twenty-eighth birthday on December 9.

Shortly before the foursome left, Spina spotted a plane circling in the distance, to the south of their location. He and the others thought it was a plane to guide the travel group toward Ice Cap Station. But when Tetley radioed the *Northland*, he was told that the plane was Pappy Turner's B-17, circling over the wrecked Duck in an unsuccessful search for signs of life.

The men at the PN9E stood in silence until Tetley said it was time to leave. Monteverde, Best, and Spina watched until they were out of sight.

THE FOUR TRAVELERS set off with Spencer out front on snowshoes, like a point man on jungle patrol. Before each step, he tapped the ground to search for ice bridges. Spencer was ideal for the job, knowing the danger they posed. Behind him, Tetley drove the motorsled, pulling O'Hara and their supplies on the attached tow sled. Wedel walked behind or alongside them on snowshoes. To play it safe, they plowed slowly for about a mile and a half through what they thought was the most heavily crevassed area. They stopped at a steep rise with an ice trough beyond it. Tetley believed that this marked the end of the crevasse field.

Tetley announced that he would gun the motorsled's engine and race up the slope, so it wouldn't stall and slide backward. He wanted Spencer and Wedel to join O'Hara on the tow sled, to spare them a difficult climb. Several yards out front, Spencer knelt to unstrap his snowshoes. Tetley climbed off the motorsled to one side, while Wedel removed his snowshoes near O'Hara and the tow sled.

Tetley told the others that he, Wedel, and Spencer would give the motorsled a hard push, after which Tetley would climb aboard. Spencer and Wedel would hop onto the trailing tow sled, like a bobsled team. Before getting started, Tetley and Wedel talked with O'Hara on the tow sled, as they waited for Spencer to join them.

Unknowingly, Tetley had parked the tow sled atop a crevasse covered by an ice bridge two feet thick, too thick for Spencer to have discovered it with his tapping and poking method. An ice bridge that thick has areas of varying strength, some able to carry weight and some not.

Spencer had walked over the bridge without incident. The motorsled had driven over it safely, and the tow sled had stopped on a solid area of the bridge. But as Wedel moved into position for the uphill charge, he stepped on a weak spot. Making matters worse, he had just removed the snowshoes that distributed his weight over a larger area.

Without warning, the ice bridge gave way, opening like a trap-door beneath Wedel's feet. He screamed and grasped for something to hold on to. Realizing what was happening, O'Hara yelled for help. He felt Wedel's mittened hands slide desperately over his legs but was unable to grab him. For an instant, Wedel gained a tenuous grip on the tow sled, but it wasn't enough. He dropped through the hole and into the waiting crevasse.

Tetley leaped onto the motorsled and drove forward to get the tow sled off the snow bridge. He and Spencer roped themselves to the motorsled and crawled on their stomachs to the edge of the hole. A short way down, they could see dark marks on a narrow ledge and more on the opposite wall. Wedel had apparently bounced from one side of the crevasse to the other on his way down. The two men stared and called into the abyss but couldn't see Wedel and couldn't tell how deep the crevasse went. It looked bottomless to Spencer.

They remained there for more than an hour, yelling for Wedel. No response. As the ranking officer, Spencer decided that they couldn't risk trying to climb down into the crevasse. There was nothing more they could do. It was time to leave.

Greenland had claimed its second victim from the B-17 PN9E crew, its tenth overall since the crash of the C-53. The death roll now read McDowell, Springer, Manahan, Everett, and Johannessen from the C-53; Pritchard, Bottoms, and Howarth from the Duck; and Demorest and Wedel from falling into crevasses.

Clarence Wedel had boarded the bomber as a passenger en route to England. He had kept the downed plane's generator working beyond all expectations. He would never celebrate his first wedding anniversary, on Christmas, less than three weeks away. He'd never meet his daughter, Reba, who'd be born the following month. As a toddler in May 1944, she'd sit on her mother's lap when her father would posthumously receive the Legion of Merit. The medal honored Wedel for "his initiative and perseverance under most difficult climatic conditions" and for displaying "a high devotion to duty and complete disregard for his own safety."

DOWNHEARTED, DON TETLEY, Harry Spencer, and Bill O'Hara discussed returning to the PN9E and taking another man to replace Wedel. But that would eat up time and require another trip across the crevasse field. Also, it might leave them short-handed for the second trip between the bomber and Ice Cap Station. Spencer decided that they should stick to the task of getting O'Hara help as soon as possible. They pressed on.

The trio moved tentatively, fearful of more hidden crevasses. The terrain was tougher, too, and they stopped frequently. Each time, Tetley killed the motorsled's engine to save gasoline. Soon, however, he had trouble starting it again. Without Wedel's mechanical wizardry, the motorsled became increasingly stubborn. The machine's lubricating oil grew thick from the cold, and soon it congealed. The oil line to the engine broke.

Tetley had been worried about the oil even before leaving the PN9E, and he'd requested that a gallon of a different grade of lubricating oil be dropped during a supply run. The oil hadn't arrived, but they'd left anyway because of the break in the weather. As he tried to fix the oil line, Tetley damaged the sled's gas line. They were about six miles northeast of the PN9E when the motorsled's engine quit altogether.

Now there were two groups of stranded men, three in the igloo under the wing of the PN9E, unaware that they were waiting for a motorsled that would never return; and three six miles away on the ice cap, one of them gravely ill and unable to move on his own. Any thought of carrying O'Hara back to the bomber was dismissed as folly. They'd stay put.

So much had happened in the month since the crash, and so much of it bad, that Spencer, Tetley, and O'Hara saw no point in bemoaning their new plight: no shelter, no radio, no walkie-talkie, no stove, few rations, and a crippled man who needed immediate aid to save his feet and perhaps his life. Plus, a blizzard was bearing down on them, a fitting start to Greenland's killing season. They focused on the lone piece of good news: they were alive. With a new storm and long hours of darkness descending, Spencer and Tetley went to work to stay that way.

First, Spencer set up a tent and carried O'Hara inside it with him. Tetley dug himself a hole in the snow and crawled in.

BACK AT THE bomber, in the ice cave under the wing, Monteverde's birthday came and went, unmarked by the celebration they'd hoped for. Without the generator, they had no light. The radio batteries grew weak. By December 11, four days after the others left, the batteries were dead, cutting their radio lifeline to Pappy Turner's supply plane and the Bluie Army bases. The walkie-talkie that Tetley had left behind was tuned to the wrong frequency, with no way to adjust it.

Clint Best was the least injured of the three, but the weeks of isolation had left him deeply depressed. Monteverde could move around, though he suffered from painful bouts of frostbite on his hands and feet. The breaks in Spina's arm, still not healed, slowed circulation in his right hand, making him susceptible to sharp aches from the merciless cold. The fingernails on his right hand had fallen off, leaving him sensitive to pain. There was little for them to do but collect supplies, tend to their injuries, and keep each other from going stir-crazy. At least they could try.

15

SHOOTING OUT THE LIGHTS

DECEMBER 1942

THE TIME HAD come for the *Northland* to leave.

Lieutenant Commander Frank Pollard acknowledged in a message to Rear Admiral "Iceberg" Smith that the ship lacked "sufficient fuel and supplies for wintering in Comanche Bay." A message sent earlier by Pollard, seeking Smith's guidance, revealed how conflicted he felt between wanting to stay and needing to go: "*Northland* desires to continue rescue operations as long as probability exists of assisting B-17 and *Northland* plane personnel." On the other hand, the message continued: "Paramount regard for *Northland* safety under present circumstances necessarily entails immediate abandonment of rescue operations because [of] inevitable risk attached to such operations. Orders are requested."

The ship almost waited too long to leave the coastline, forcing it to break through a five-mile-wide belt of pack ice to reach open water. Once there, the *Northland* was out of range of radio communications from the men on the ice or at the army's bases and stations, ending the ship's direct involvement in the rescue efforts. Yet the *Northland* left a great deal in its wake.

Still ashore were the remains of the rescue team of Pritchard and Bottoms, as well as their passenger, Howarth, and also the wreckage of the Duck. Also left behind were five members of the *Northland*'s crew, led by an intrepid twenty-two-year-old ensign named Richard Fuller. The Coast Guardsmen under Fuller's command, all fellow volunteers, went ashore by boat at Beach Head Station on December 4. They hoped to help Monteverde's PN9E crew and recover the bodies from the Duck. They might also have looked for McDowell's C-53, but the cargo plane remained lost, likely buried under snow with the bodies of its crew.

Fuller and his team made several valiant attempts to reach the B-17 and the Duck, but were unable to reach either plane. It wasn't for lack of trying. The rescue effort, supposed to last no more than two weeks, turned into a five-month ordeal. Over the winter of 1942–1943, much of their time was spent trapped at Beach Head Station, a wooden hut described by Chief Pharmacist's Mate Gerard Hearn as "an overgrown crate, about thirty feet square." With Fuller and Hearn were Stanley Preble, a seaman; Harold Green, a fireman; and Donald Drisko, a mechanic.

Fuller and dogsledder Johan Johansen also holed up at Ice Cap Station during the rescue attempts. When the stove vents there filled with snow, they lived in fear of death by carbon monoxide poisoning. Their kerosene ran out, and they spent long stretches in the dark or in flickering candlelight. Over time and repeated blizzards, the flat-roofed shack was buried in snow. Fuller suffered a frostbitten foot, and three of his toes turned black, though he later recovered. Their radio died, two inches of water pooled on the floor, and nine dogs on their sled team froze to death. The men shared their quarters with the remaining six dogs, whose wastes turned the station into a reeking kennel.

They spent days tucked in their bunks for warmth. They emerged to play cards by the light of a single candle, or to use a snow tunnel they'd carved for a latrine. Ice Cap Station was eventually deemed unfit for human habitation and they rejoined the other men at Beach Head Station. Conditions were little better there, a sixteen-by-twenty-four-foot shack so covered by snow that their only access was through an attic loft window.

By the time they were picked up the following spring, the Coast Guardsmen had spent more than five months in conditions hardly better than those of the men they'd hoped to help. Perhaps most frustrating, they initially were given the wrong coordinates to search for the Duck; even when the location was corrected, Fuller's team wasn't told. Nevertheless, Fuller received the Navy and Marine Corps Medal, and all five received commendations for "courage, energetic and cheerful cooperation, and devotion to duty." The official Coast Guard history of the war gently acknowledged that they never found the downed air crews: "This expedition had to be evaluated more in terms of heroism than accomplishment."

MEMBERS OF A RESCUE TEAM STAND ON THE ROOF OF SNOW-COVERED BEACH HEAD STATION. (U.S. COAST GUARD PHOTOGRAPH.)

BECAUSE OF DWINDLING daylight hours, Pappy Turner and the five men of his B-17 crew relocated from Greenland's west coast to the unfinished base at Bluie East Two. There, they'd be less than 150 miles from the downed PN9E, minimizing nighttime flights from one coast to the other. Storms grounded Turner's B-17 for two days after he spotted the downed Duck, but he was able to get his bomber back into the air on December 9.

By a stroke of good fortune, Turner and his crew spotted Spencer, Tetley, and O'Hara at what became known as the Motorsled Camp. With no way to communicate, the men on the ice couldn't tell Turner's B-17 what had happened to Wedel, so Turner and his crew didn't know who or how many men they were helping. Among the supplies they dropped was the motor oil that Tetley had requested days earlier. Spencer and Tetley worked for days on the motorsled but couldn't restart the engine. They abandoned it, and soon the machine was buried under several feet of snow.

Indeed, more snow was the one thing they could count on. Drifts piled up so high that O'Hara's side of the small tent threatened to collapse and bury him alive. Spencer and Tetley spent that night taking turns shoveling it away from the canvas. When morning came, Spencer announced that they needed to prepare for the long haul. That dealt a blow to Tetley's spirits, and he remained cooped up in his snow hole for several days.

In the meantime, Spencer, with a little help from O'Hara, dug an ice hole they could use for cooking. Then they dug an adjacent hole about three feet deep, with floor space about six feet by nine feet, to sleep and pass the days. With nothing else to do, they burrowed deep enough to create a six-foot ceiling in their ice den where Spencer could stand and stretch.

Tetley emerged from his funk and dug a passage from his hole to Spencer and O'Hara's, the start of what turned into a warren of connected holes in the glacier. On the surface of the ice cap, they built a wall of snow blocks around a tunnel-like entrance and covered it with ice-encrusted sleeping bags, which served like the flap of a tent. They cooked beneath the entrance, so the heat from their stove wouldn't melt the roof of their cave and send icy rivulets pouring onto their sleeping area.

Nighttime snowfalls drove down into the entrances, so Spencer kept his shovel with him to dig out every morning. Then he'd go to Tetley's hole and dig him out, too. They expanded their quarters again, arranging their skis like an A-frame hut over a new opening to their subglacial home. The snow piled up around the skis, and the men turned the frozen tepee into a cold storage room for rations and other supplies.

Pappy Turner's crew dropped provisions whenever possible, but the Motorsled Camp men couldn't always collect them. One day, with two K rations remaining, they decided to eat everything and take their chances until the next drop. Their stove was unreliable, so Tetley babied it to keep the flame alive. But hypothermia made him sluggish, and as he warmed their last meal he knocked over the stove, spilling their rations into a nasty mixture of snow and gasoline. They ate what they could and made coffee, but then that spilled, too. Fortunately, Turner returned the next day with fresh rations. Their food supply ran low again as Christmas approached, but the Motorsled Camp crew ignored the risk. They ate full shares, sang carols, and tried to make the best of it. Pappy Turner's B-17 returned three days later to restock their storehouse.

They'd found a way to survive, but O'Hara's feet continued to get worse. A bout of diarrhea cost him more weight, and he was often sluggish. Yet he held on without complaint. Back when their B-17 first crashed and O'Hara could go outside, he marveled at how the night sky glowed with the aurora borealis. But as weeks of misery dragged on without end, the northern lights seemed to taunt him with their liquid beauty. O'Hara dreamed of shooting them from the sky.

IN THE SNOW cave beneath the PN9E's right wing, Monteverde, Spina, and Best settled into their own routine.

Much of their day, and much of their energy, revolved around making trips outside to collect supplies dropped by Turner's B-17. Inside their igloo, they tried to be creative with their rations, at one point using chocolate and malted milk to make snow-based ice cream. They improvised a recipe for fudge, too.

The trio lived every moment with the pain of being wet to the skin and cold to the bone, of weakened muscles that ached from shivering, of stiffened joints locked like rusted machinery. Candles that Turner dropped rarely lasted long, making the twenty-hour Arctic nights seem even longer. During storms, entire days passed when they didn't see light. Like the men at the Motorsled Camp, Monteverde, Spina, and Best had no working radio or walkie-talkie, so they couldn't communicate with anyone but each other. They couldn't ask for items they wanted or needed, and they couldn't enjoy the comforting sound of a voice, or even a coded message, from beyond their frozen room. Turner and his B-17 supply crew weren't even certain that all three men were still alive. When they flew overhead, they might see one or two emerge from under the wing to collect the dropped packages. They could only hope that the third was resting inside.

Monteverde and Spina struggled but bore up under the deprivations, the boredom, and the stress. But Clint Best's mind bent under the strain.

FROM HIS POST at Bluie West Eight, Colonel Bernt Balchen closely tracked the failed efforts to reach the stranded men by land. On December 1, he wrote in his log, two dogsled teams left Beach Head Station for Ice Cap Station, intending to go from there to the PN9E. But they turned back because an army lieutenant leading one of the teams couldn't control his dogs. Two days later, another search team left Ice Cap Station but returned because they "saw lights moving toward station [and] decided Tetley had returned." They were mistaken. Another attempt began four days later, but returned as a result of bad weather and rough terrain. Three dogs died and one ran off during that effort. On and on it went, with dogsleds and motorsleds breaking down or bogging down; dogs running off or dying; men suffering from frozen feet; and storms making travel and navigation impossible.

As days stretched into weeks, the inability to retrieve Tetley and the five remaining survivors of the PN9E crash stirred worry, frustration, and embarrassment not only in Greenland but throughout the military. Brainstorming about possible ways to bring the men home reached the highest levels of the U.S. Army and Navy, though at least some ideas reflected a lack of understanding about the severe conditions on Greenland's ice cap.

Military planners discussed using helicopters, not realizing that storms would spin the whirlybirds like tops before smashing them to pieces. Another idea proposed by army leaders was to drop large cargo gliders onto the ice. Under that plan, the six men would climb aboard, and then low-flying planes would snatch the gliders back into the air with hooks hanging from their bellies. As crazy as it sounded, the idea was only half nuts. In fact, the Army Air Forces would employ a glider drop-and-snatch scheme in June 1945 in Dutch New Guinea. The targets of that rescue were three plane crash survivors, one a beautiful member of the Women's Army Corps, who were stranded among Stone Age tribesmen in a remote valley known as Shangri-La.

"Has Army considered use of auto-gyro or helicopter as means of rescuing personnel in Greenland?" the navy's commander in chief of the Atlantic Fleet inquired. Two hours later, a reply came from Admiral Ernest J. King, the navy's overall commander in chief: "Army has considered use of auto-gyro helicopter and gliders, but has rejected their use as impracticable under existing high-wind conditions."

None of these discussions were known or even hinted at outside government and military circles. Newspaper reporters and radio correspondents were covering every aspect of the war, and journalists would have salivated at the prospect of telling stories of multiple Greenland plane crashes and heroic rescue attempts. The ongoing drama of six servicemen trapped in ice caves six miles apart would have been like catnip to battle-weary newsmen and newswomen.

But all war-related events in and around Greenland were Allied military secrets, and no stories leaked into newspapers or onto the airwaves. If the Nazis learned from news reports about a B-17 bomber lost on the ice, the thinking went, they might try to find it, kill its crew, and steal its Norden Bombsight. Or, if the enemy knew that the *Northland* was anchored in Comanche Bay, the ship would have made an appetizing torpedo target for a U-Boat.

Even when family members were told that their husbands, sons, or brothers were missing or killed on the ice cap, they were instructed not to share any details until the military made the news public. Loose lips sink ships, they were told, and they listened.

The six men on the ice had no idea what, if anything, their loved ones knew of their plight. But they understood the rules of war and the larger forces at work. Their job was to stay alive long enough to explain why they'd stopped writing letters home.

16

SNUBLEBLUSS

MARCH–MAY 2012

Before leaving the gloomy January 2012 meeting at Coast Guard Headquarters, I sought out Commander Jim Blow, the Coast Guard's point man on the Duck Hunt. I expected Blow to confirm that we'd just witnessed a bureaucratic waterboarding.

"Now what?" I asked, as plaintive as Lou had sounded in the meeting.

When Blow answered, I suspected that he'd nodded off during the dour, you're-on-your-own message delivered by Lieutenant Colonel James McDonough from the Defense Department's missing-in-action office. But Blow is career Coast Guard, a man used to doing more with less. Also, he has the steady pulse of a rescue pilot. Harsh words in a boardroom don't faze him.

He looked at me as though I'd asked a foolish question. "We'll need a detailed mission plan," Blow said. "This isn't over. We're proceeding as if we're going."

I nodded and left, unsure whether to feel confused, calmed, or both.

But that was two months ago. Since then, wheels have spun,

calls have gone out, and hundreds of e-mails have flown, yet progress has ground to a halt. Lou's plan to go to Greenland in May is
off; it will happen in August, at earliest. As far as I'm concerned,
the Duck Hunt is in peril.

THE COAST GUARD remains open to providing a huge C-130
cargo plane, but by all indications there's no money in the service's budget for more than that. Lou's been seeking contributions from private supporters, but that hasn't panned out, either.
Everyone likes the idea of bringing home the remains of three
brave American airmen trapped under the ice since 1942, but no
one is willing to pay for it. No one, apparently, except me.

Using the advance payment from my publisher to write this
book, I've written a check to Lou's expedition company, North
South Polar. Ostensibly, it's a loan that guarantees me a seat on a
trip to Greenland. Lou promises to return the money by May 1 or
to apply it to the cost of my travel and provisions and refund the
rest. But May 1 passes, the money remains in Lou's hands, and no
travel date is set. I'd like to think that my money has been sitting
in a bank account, but I'm not that naive. Lou's been floating this
mission for two years, exhausting his savings as well as himself,
and I'm certain that the money was used to keep things afloat. I
could ask Lou about it, but I'm not sure I want to know.

In the hope of raising significant sums, Lou's been relying on
a Los Angeles–based producer named Aaron Bennet. He's been
pitching television networks on a reality/adventure show based
on searches for missing airplanes and lost airmen, with Lou and
North South Polar as the stars. Lou has a whole series in mind,
with missions not only in Greenland but everywhere from Antarctica to the South Pacific, from planes lost in the 1930s to Vietnam-
era wrecks. The Duck Hunt is first on the list. Bennet believes in
Lou, and I can tell from e-mails and phone calls that he's working
hard. But so far he's only received what I call "Hollywood yeses."

They're better than no's, but they're really maybes, which makes them the contractual equivalent of air kisses.

Television networks are overwhelmed by pitches for reality shows featuring daring adventurers, as well as shows about celebrity lifestyles and dysfunctional housemates. Another problem facing Bennet's pitch is more basic. It's the same issue that troubled the military's missing-in-action experts: no one is certain where to find the Duck. Like the Defense Department, Hollywood wants a sure thing.

AT FIRST BLUSH, the Duck's resting place might seem relatively easy to pinpoint. In the months after the crash, the wreckage was spotted multiple times on the ice. Those sightings all but eliminated any possibility that the Duck plunged into the water and sank to the bottom of Koge Bay. The last confirmed sighting was in 1947, five years after the crash. That's good news, because it meant that the Duck wasn't on a fast-moving glacier carrying it swiftly toward the bay.

Also significant were the hand-drawn historical maps that witnesses made after their sightings, especially two by Bernt Balchen, each with an X marking the spot where he saw the Duck. Lou has grown spellbound by Balchen's maps, the original version torn from Balchen's notebook as well as a later one that Balchen did in watercolor. By comparing radar data and satellite imagery with these maps, Lou has concluded that geographic features long dismissed as fanciful flourishes correspond to real parts of the landscape. If so, they represent potential landmarks for the Duck Hunt.

Lou has also been studying a topographical map by Herbert Kurz, the navigator on Pappy Turner's B-17, who marked a similar spot with the Duck's location. The maps made by Balchen and Kurz tell roughly the same story, and both fit the known facts about the Duck's disappearance. For instance, they place the

Duck's wreckage in a spot about halfway between the PN9E and Comanche Bay, where the *Northland* waited in vain. The location matches evidence that Pritchard radioed the ship for help—his requests for magnetic orientation—about nine minutes into what should have been a twenty-minute flight.

If the Duck had crashed in a South Pacific jungle or in a European forest, searchers likely would have more than enough in-

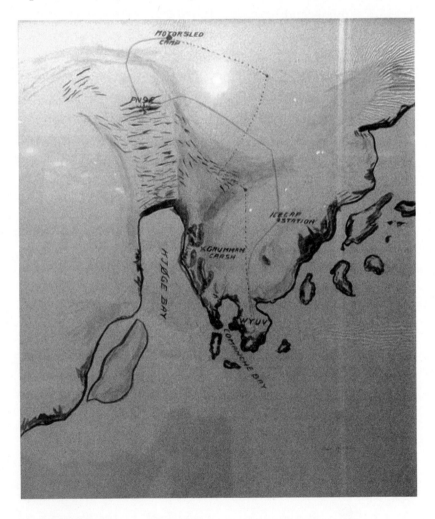

WATERCOLOR VERSION OF BERNT BALCHEN'S MAP OF THE KOGE BAY AREA. *(U.S. COAST GUARD IMAGE.)*

formation to find the wreckage. Not in Greenland. The search for the Duck is complicated by three main factors: accumulating snow, glacial movement, and a mixture of errors and conflicting accounts in official reports about where it went down.

First, despite recent melting of Greenland's ice, during the seventy years since the crash some thirty or more feet of snow and ice might have built up atop the Duck. That means finding it will require ground- and ice-penetrating radar, followed by drilling, digging, or melting for confirmation. Second, despite the 1947 sighting, it's not certain that the Duck wasn't on an active glacier moving toward Koge Bay. On their trip to Greenland in 2010, Lou and his team left behind tracking devices that suggest the ice in the area is barely moving. But if that's not the case, the Duck would be nowhere near where it crashed. At some point since November 1942, it might have splashed into the bay with a newly calved iceberg.

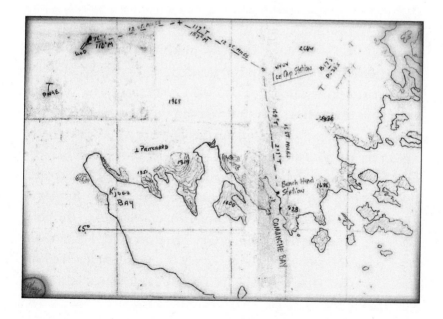

MAP BY LIEUTENANT HERBERT KURZ, SHOWING THE CRASH SITES OF THE PN9E AND THE DUCK. *(U.S. COAST GUARD IMAGE.)*

The third factor is the conflict in historical records. The Coast Guard's John Long and other researchers have found nearly a dozen sets of reported latitude and longitude coordinates for the crash site. Some were clearly erroneous and were later corrected, but even the ones considered credible are inconsistent. Most are on a tongue of land on the east side of Koge Bay, but when plotted on a map they make a shotgun pattern. For example, the military's official PN9E accident report from April 1943 places the Duck's wreckage at the intersection of latitude 65 degrees 8 minutes north, and longitude 41 degrees 0 minutes west. That's more than two kilometers, or one and a quarter miles, from the coordinates that Pappy Turner gave to Colonel Balchen in December 1942. Some points are even farther apart, but those two locations are considered among the most credible by Duck Hunt historians, notably John Long and retired Coast Guard captain Donald Taub. On the other hand, the accident report from April 1943 was focused on the PN9E, so there's lingering skepticism about how precisely it places the Duck crash. And even if the coordinates were once correct, there's no telling how much the Duck might have moved since then along with the glacier.

To narrow the search, Lou and the Coast Guard have collected radar data from planes that have flown over the area on scientific missions. Among their sources are the University of Kansas–based Center for the Remote Sensing of Ice Sheets, known as CReSIS; a U.S. Navy plane with advanced radar equipment that was returning from Iraq and Afghanistan; and a NASA mission known as Operation IceBridge that's collecting data on changes to the polar ice sheets. In addition, while in Greenland in 2010, Lou and his team did a boots-on-the-ground survey of one radar-identified site that the Coast Guard considered promising. It was a false lead; Lou thinks that a radar image resembling a plane was in fact meltwater that pooled atop bedrock under the ice.

Lou has spent countless hours this year combing through

the historical coordinates and the radar data, and he's become a proficient computer-aided cartographer. After plotting and cross-referencing all the potential locations on two- and three-dimensional maps, Lou declared in a memo to the Coast Guard that he's reached "a very high degree of certitude on the [Duck's] current location." Lou's target coordinates are within five miles, or less than eight kilometers, of the official PN9E crash report, and even closer to Turner's December 1942 sighting. Privately, though, Lou knows that even the best use of radar and historical data is educated guesswork. The only way to be sure of the Duck's location is to burrow inside the glacier and obtain hard evidence such as a photograph.

But time is slipping by, and the weather makes late spring and summer the only practical times to search Greenland for missing planes. If a green light isn't lit soon, a year or more will pass before we step onto the ice. Momentum might wane and money might get even tighter, increasing the likelihood that the Duck and its men will be lost forever.

This prospect worries me. I'm determined to get to Koge Bay this summer, to walk the glacier in my own boots, to see the area where the men of the PN9E holed up and where the Duck went down. I've begun contacting private expedition companies on the east coast of Greenland, inquiring what it would take to mount a micro-mission on my own. I'm the first to admit that this would be far from ideal. I'm up for a challenge, but my experience in extreme cold consists of shoveling snow from my driveway. My poor directional sense is a source of humor for family and friends. It's possible that I'm neither as young nor as fit as I think.

When I set aside romantic fantasies of finding the Duck, I know that I'd be lucky just to avoid falling into a crevasse, getting lost in a storm, or upsetting a hungry polar bear. I'm also aware that going alone means that I wouldn't have access to advanced ground-penetrating radar, and I wouldn't have the equipment or the necessary permits to dig deep enough to rule out or confirm possible sites. I'd be relying on luck, on a Greenland guide whom I've never met, and on the remote possibility that recent melting had exposed part of the wrecked plane. By comparison, Lou seems like a model of moderation. But I don't see any other choice. One way or another, I'm going to Greenland.

Meanwhile, Lou's been picking up odd jobs to stay afloat. One night he e-mails me from a work site, asking for a historical document I promised to send him: "I've got the computer out on top of a Dumpster working away! So shoot me anything you have. I'll be here for a few hours." The image of Lou using a trash bin as a desk fills me with a nauseating mix of despair and admiration. I know how hard he's trying, how much he's put into this, but outside sponsors still haven't signed on; no television network has committed to film the expedition; and time is growing short. Lou remains optimistic—"Don't worry, we're going" is his new favorite line—but from our daily e-mails and phone calls, I know that he's feeling the stress.

Worried that Lou is about to delay the trip, I put my backup plan in motion. I choose a guide; select commercial flights to the Kulusuk Airport on Greenland's east coast; buy glacier glasses and new snow pants; and, with my guide's help, tentatively hire a speedboat captain for a ride to Koge Bay.

JUNE–AUGUST 2012

Just when I'd abandoned hope for Lou's mission, there's an unexpected turn of events. Leading with his heart and an almost religious fervor, Lou has gained traction. A big moment comes when Lou tells me that the Coast Guard has lined up a C-130 for the expedition, with plans to depart from a small airport in Trenton, New Jersey, on August 20.

Excited, I telephone Commander Jim Blow to confirm the good

news. When I reach him, Blow is in the middle of arranging helicopters to take us from Kulusuk to Koge Bay. He's too much of
an officer and a gentleman to say so, but I swear that I can read his
mind: "I told you we're going." He says final approval for the C-130
is still pending, but he feels good about our chances and tells me to
plan for a week or more of glacier camping.

With a departure date in hand, Lou's groundwork pays off.
Government officials in Greenland issue an elaborate fifteen-page
permit for North South Polar to search out not only the Duck but
also McDowell's C-53 and planes left behind from the Lost Squadron. Corporate sponsors step up with equipment and supplies. Lou
slashes his original million-dollar budget by more than half, to
the narrowest margin that assures a safe and well-provisioned mission. Lou's revised plan focuses purely on finding the Duck, with a
recovery mission to follow, if and when the plane and its men are
located. "If we don't get onto the ice we have nothing," Lou tells
me. "Once we get there and we're successful, everything else will
follow."

I cancel my plan to go it alone and send more money to Lou's
nonprofit Fallen American Veterans Foundation. Three weeks
later, I send even more.

Lou's team starts to congeal, bringing expertise in fields ranging from geophysics to radar to mountaineering to excavation.
They trade flurries of e-mails on everything from bedrock depths
to the warmest sleeping bags to the best method of polar bear
deterrence. This last leads to a lengthy discussion of weaponry,
electrified fences, and something called a "Snublebluss," a tripwire
that activates an alarm and warning lights around a campsite. I
like the name but worry about its effectiveness. One e-mail includes grisly photos of a polar bear attack victim.

In the midst of these discussions, I pull from my files a government pamphlet titled *Encounters with Wildlife in Greenland*. I
highlight a long section on mortal dangers: "Despite its size and

awesome strength, the polar bear is swift and agile, moves easily on rough ice and steep slopes, and is an excellent swimmer. . . . Polar bears are meat eaters . . . [and] any animal, including humans, is potential prey." The recommended response is avoidance, and the guide offers instructions to make a "Chili-Con-Carne Alarm," using a can of strong-smelling meat stew as bait to trigger a siren. Call it a Chili-Con-Snublebluss. If escape or scaring the bear isn't possible, hope that a gun is handy. "Avoid head shots," the guide cautions, "as they often do not kill a bear. Do not check the results of your shot. If the bear goes down, keep shooting vital areas until it is still. Make sure it is dead." Noted.

When I show Lou the guide, he assures me that the chance of encountering a polar bear is almost zero. Still, Lou has recruited a medical student and U.S. Army National Guard captain just back from Afghanistan named Frank Marley to be the expedition's chief of health and security.

Yet for all the fears of polar bears and hidden crevasses, of vicious storms and killing cold, our biggest worry can't be overcome with Snubleblusses, electrified fences, safety ropes, guns, or extreme weather gear. Hovering over every conversation, every e-mail, every decision, is the nagging anxiety that even if we reach Koge Bay, we won't find the Duck.

In July, after the team is assembled and a mission plan is written, Lou sends a triumphant text message at one in the morning: "Everything is a go." We're bound for Koge Bay with radar-generated search coordinates, a historic treasure map, an expert team, the Coast Guard's support, and Greenland's approval, for what might be the last chance to find John Pritchard, Ben Bottoms, Loren Howarth, and the Duck.

By phone the next day, I admit to Lou how deeply I'd doubted him. "Yeah, I figured," he says, laughing. "No sweat, man. We'll just leave you on the ice."

But during the weeks that follow, the mission again teeters on the edge of oblivion. Money remains the main sticking point, as hopes for a television deal fizzle and counted-on sponsors come up short. Stressed, I unload on Lou and his producer/partner, Aaron Bennet, for not having everything in place, even as I send Lou more cash and use my credit card to pay the balance of a bill for sleeping bags. A few days later, I give Lou my credit card number so he can buy tents, a rifle and a shotgun, boots, gear, and assorted other equipment, with an understanding that I'll be repaid when other money arrives. Soon I'm answering antifraud calls from American Express, which apparently wants to be sure my account hasn't been hijacked by a mad survivalist.

After thanking me for the card, Lou asks, "What's my limit, Dad?" We both laugh, me more ruefully than him. In no time, Lou blows past the limit I set.

FORTUNATELY, I'M NOT the mission's only potential funder. In addition to the C-130, the Coast Guard appears ready to provide as much as $150,000, as "support for expedition to Greenland to provide positive location of USCG J2F-4 Grumman Duck suspected crash site." A document seeking bids for the job talks about "positive location," but if the contract comes through, Lou and North South Polar will only be expected to investigate—and either confirm or rule out—six "Points of Interest" considered the most promising from radar hits and historical research. If those sites don't pan out, up to four more locations might be examined, time permitting. Effort is guaranteed, success isn't. Yet what's happening is clear: at the urging of Jim Blow, the Coast Guard is getting ready to close the major gaps in the expedition's streamlined budget.

It's safe to say that there could be no odder couple than the Coast Guard and North South Polar. If the contract is consummated, it would be a marriage of discipline and dreams, and the

Duck Hunt's civilian and military point men personify the contrast. While Lou sallies forth to slay dragons, Jim Blow is a study in precise control and military planning, as rooted in reality as his regular haircut appointment.

AT FORTY-FOUR, JIM is married to his college girlfriend, a nurse in a neonatal intensive care unit. They live in suburban Virginia with their three sons, at least two of whom can imagine becoming pilots like their father. He's spent nearly twenty years in the Coast Guard, and the service defines him. His father was a navy flier, but Jim prefers rescue work. "The navy is always training for something that might never happen. With the Coast Guard, you're training for what you do day in and day out. Making that rescue, making an impact on people's lives."

COMMANDER JAMES "JIM" BLOW. *(MITCHELL ZUCKOFF PHOTOGRAPH.)*

The lineage from rescue flier John Pritchard to Jim Blow is easy to trace. One of Jim's best days at work came when he was flying a twin-engine Falcon jet, searching for a missing diver in the Gulf of Mexico. "Those searches usually don't end well," he says. He was assigned to fly a search pattern a maximum of four miles from where the diver was last seen. Jim decided that wasn't far enough, so he stretched the area to seven miles. When turning his jet at the far edge of the enlarged area, he spotted the missing diver from his left-side window. Soon the man was safe and dry.

To win Coast Guard funding to find Pritchard, Bottoms, and Howarth, Jim has written his superiors a lengthy brief titled "Operation Duck Hunt 2012," with justifications including the military ethos of "leave no man behind"; the risk of climate change exposing the crew's remains to wildlife; the rising value of World War II aircraft luring unscrupulous salvagers; and Nancy Pritchard Morgan Krause's advancing age.

Unknown to Lou, Jim also arranges for the U.S. Army's Cold Regions Research and Engineering Laboratory, known as CRREL, to send a radar team to Greenland. His reputation at stake, Jim wants reconfirmation that something might be found under the ice near Koge Bay. While Jim awaits those results, his superiors approve his "Operation Duck Hunt" request and authorize spending up to $150,000 on the mission.

But then Lou submits a bid seeking almost $200,000, and the mission flirts with disaster. For Lou, the bid reflects something closer to his true costs, with all the technology needed to do the job. Even at that price, he says, he would take significant loss, after working unpaid on the mission for two-plus years. Having seen how Lou operates, from a trailer in his yard as an office, I don't doubt that money is secondary. But the government can't pay Lou for the work he's done on his own, only the work being requested on the ice. Jim Blow recoils at the high bid.

"We found this money," he says angrily, "but he's asking for more, without justifying why." Jim won't ask his superiors for more, and in mid-August he sits at his desk at Coast Guard Headquarters ready to scrub the mission. He's spent countless hours studying charts and photographs, puzzling over Balchen's treasure map, and imagining himself in the Duck's cockpit, to calculate where Pritchard might have gone wrong. Still, he can't defend a wild goose chase, and he worries that Lou doesn't have his act together.

Before Jim can bring himself to abandon the mission, the CRREL folks call with results of the radar survey. More than half a dozen locations inside the glacier reveal "anomalies," potential metal targets. Several are labeled "strongly prospective targets."

In quick succession, a chastened Lou lowers his bid to $150,000, the contract is approved, and Jim commits himself and the Coast Guard to the expedition.

"That radar report gave me the warm and fuzzy feeling I needed. That report probably saved this mission," he says. Jim repeats a line I've heard from Lou: "If it's there, we're going to find it. If we don't, it's not there."

Lou says that he, too, was close to abandoning the expedition rather than being driven deeper into debt. He says he went ahead because he was determined to complete the mission. "It's never been about the money," he says. "We—the families, the Coast Guard, and all those who put in so much time, effort, and energy—have come too far to let it all end here. So I bit the bullet."

WHEN THE LAST major hurdles are cleared, Lou and I agree to pack a bottle of Scotch. We'll toast either to our success or to the respectable failure of having given our all. But not just any whiskey will do. Lou likes the idea of drinking the same liquid fortitude that explorer Ernest Shackleton hauled to Antarctica in

1907 during a failed attempt to reach the South Pole. Shackleton left behind three cases, which were found buried in permafrost in 2007. A Scottish distillery replicated the blend and now sells it for $170 a bottle.

Apparently I'm buying.

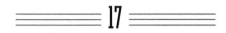

17

OUTWITTING THE ARCTIC

DECEMBER 1942–JANUARY 1943

EACH NEW DAY seemed to bring a new rescue attempt for Don Tetley and the remaining five PN9E crewmen, without a resulting rescue.

In mid-December 1942, U.S. military officials turned to a Canadian bush pilot named Jimmie Wade, who volunteered to land a twin-engine plane fitted with skis on the ice near the Motorsled Camp. The plan, almost as daring as Pritchard and Bottoms's landings in the Duck, called for Wade to pick up the three men there and fly them about 140 miles to the base at Bluie East Two. Then he'd return and do the same for the three men in the snow cave under the PN9E wing.

Wade was a civilian pilot for a private Canadian airline, Maritime Central Airways, but his bosses agreed to loan him to the U.S. military for the rescue effort. Along with Wade's services, the airline leased the United States government a sleek and sturdy skiplane called a Barkley-Grow T8P-1. Wade would be the pilot, and a U.S. Army captain named J. G. Moe from the Air Transport Command would serve as navigator.

On December 22, Wade and Moe took off from Bluie West One, heading across Greenland toward Bluie East Two, where they intended to refuel before making the short hop to the stranded men. The weather reports were good, but the weather itself wasn't.

The flight was at the outer reach of the ski-plane's range, and Wade and Moe ran into powerful headwinds that forced them to burn more fuel than expected. Unfamiliar with the island's jagged coastline and hampered by fog, Wade turned into the wrong fjord, thinking that it was the route to Bluie East Two. Before Wade and Moe could correct their flight path, they ran perilously low on fuel. Down to their last fumes, Wade steered toward what he and Moe thought looked like a solid stretch of snow-covered sea ice. They were wrong. Upon touching down, the Barkley-Grow sank through the thin ice. Wade and Moe grabbed whatever equipment they could reach and abandoned ship.

Freezing and soaked, the two men spent several days in a rubber dinghy, slogging across, over, and through a mile of treacherous fjord ice to the coast. Travel on land was only slightly easier. They walked, leaning forward with their heads bowed, into grainy snow driven by fierce winds. A week after Wade and Moe went down, their supplies dwindled to two Fig Newton cookies a day. Rescue planners gave them up for dead.

When hope seemed lost, the downed fliers stumbled upon a band of Inuit hunters. Just as answering an SOS call is the law of the sea, offering aid to lost travelers is the rule of the Arctic. The hunters brought the strangers to their village and nursed them for several days. On January 2, eleven days after their flight, Wade and Moe were delivered by native dogsled to Bluie East Two, where they remained until late spring. Six months after being rescued, Wade received the British Explorer Medal, another example of a would-be rescuer honored for a brave but unsuccessful attempt to help the PN9E crew.

THE END OF Wade and Moe's mission came the same day that another dogsled team was forced back to Beach Head Station by deep snow. With that latest failure, American military leaders in Greenland reached a breaking point.

Their priority remained the lives of six men on the ice cap, but their own reputations were on the line, too. Military and civilian officials in Washington were being briefed regularly about the rescue efforts. Questions might soon arise about the competence of the men running the war in Greenland.

Five days after Wade and Moe went down, Colonel Bernt Balchen was summoned to Bluie West One from his remote northern headquarters at Bluie West Eight. Aides ushered him into a meeting with the military's top land and sea officers on the island: Rear Admiral Edward "Iceberg" Smith of the Greenland Patrol and Colonel Robert Wimsatt, commander of the U.S. Army's Greenland bases. This was the same Colonel Wimsatt whom Balchen had helped to rescue months earlier. When Smith and Wimsatt asked him for suggestions, Balchen outlined what he considered a surefire plan to rescue Don Tetley and the PN9E crew. All he needed was their support and a couple of very valuable airplanes to pull it off.

Balchen's plan demonstrated why he was a rare and talented airman. Plenty of pilots were fearless or seemed so, but few could match his ability to synthesize experience and knowledge to maximize the potential of flying machines. Balchen told Smith and Wimsatt that he wanted to apply the lessons of John Pritchard's two glacier landings and takeoffs in the Duck, and then combine those feats with a stunt borrowed from the annals of Arctic exploration.

Despite the Duck's crash, Pritchard had demonstrated that it was possible for an amphibious plane to use the Greenland ice cap as a belly-down runway. Balchen told his bosses that a much larger seaplane could do the same thing, with even greater effectiveness. As proof, he cited the first attempt to reach the North Pole by airplane. Seventeen years earlier, in 1925, Roald Amundsen and Lincoln Ellsworth attempted to fly two seaplanes to the top of the world. The effort failed, but not before two pilots hired by the explorers took off in large, heavily loaded seaplanes by skidding them across the ice of King's Bay in Norway. Balchen knew the details firsthand because he participated in rescue efforts for the expedition as a young flier in the Norwegian Air Service.

As he built the case for a belly-down-on-the-ice rescue effort using large amphibious planes, Balchen also had a more recent example. Six months earlier, in June 1942, U.S. Navy pilot Dick Parunak had landed an amphibious plane belly-down on a temporary lake on Greenland's ice cap, as part of the Balchen-led *My Gal Sal* B-17 rescue. Now Balchen wanted to try what might be called a modified Pritchard-Amundsen-Ellsworth gambit, with a Parunak twist.

Balchen described his plan as "one last trick to outwit the Arctic." He proposed using a bigger plane than Pritchard's Duck. Greenland's ice cap would serve as a substitute for Amundsen and Ellsworth's frozen Norwegian bay and Parunak's temporary lake. Balchen told Smith and Wimsatt that his airplane of choice would be the navy PBY-5A Catalina flying boat, the same model plane that Parunak flew.

Balchen had good taste in flying ships. The plane he'd chosen was a marvel. PB stood for "patrol bomber," and Y was the letter assigned to its manufacturer, Consolidated Aircraft Corporation. The PBY Catalina was the workhorse of naval aviation, a rugged amphibian with a range of 2,500 miles. Some four thousand would be built before the war ended. They could drop bombs on U-boats on the way into battle, and could rescue drowning sailors on the way out. On bombing runs, crews called the PBY Catalina "The Cat." But on rescue missions, it was affectionately called Dumbo, a tribute to Disney's flying elephant, which lit up movie screens in October 1941. At almost 64 feet long, with a wingspan of 104 feet, the Dumbo dwarfed the Duck.

Wimsatt liked the idea, but Smith balked. In the admiral's view, Balchen's plan was too dangerous. Only four PBY Catalinas were in service in Greenland, and they were being used to locate and harass U-boats attacking Allied merchant ships during the ongoing Battle of the Atlantic. Smith feared diverting half his PBY fleet to a dangerous and untested rescue attempt.

Danger to crews and equipment was a legitimate concern. A PBY Catalina might suffer catastrophic stress and break into pieces during a hard landing on snow and ice. Smith also knew that the most challenging part of the rescue might not be the belly landing, but the very act of flying over Greenland in the midst of winter. Pritchard and the Duck had done fine in landings and takeoffs, but they went down as a result of storms and fog. In fact, Smith had already squashed discussion of sending the Coast Guard cutter *North Star* close enough to the east coast to use its Grumman Duck for a rescue attempt. He didn't want to lose more men and planes to Greenland's weather.

Smith countered with a more conservative approach. He suggested that a new motorsled be flown to Greenland from the United States and be dropped by parachute to the men at the Motorsled Camp, so they might continue on their own toward Ice Cap Station. In the meantime, he said, Pappy Turner's B-17 would continue to drop supplies whenever possible.

The meeting ended in a stalemate. Balchen wrote later that he vowed, "If I'm to crawl in on my hands and knees, I'll get the boys off the Ice Cap." Balchen described Smith's reaction as "a glacier-cold shoulder." He added sarcastically: "No planes for me for such a lunatic purpose."

To bolster his easy-does-it approach and cover his tracks, Smith enlisted the support of his superiors. He explained in a message that Balchen "desires . . . [PBY] to land on Ice Cap, which I have informed him is considered too great a risk at this time of year. [I] believe other possibilities have not been exhausted." Smith's message added, however, that if ordered to provide a PBY he would do so, and he acknowledged that Balchen should oversee such an operation because he was the "most experienced Arctic flier now here."

Initially, the navy brass supported Smith's position. One reply from Smith's superiors echoed his position, declaring that "aircraft rescue missions are warranted only in the event such operations do not unduly hazard the aircraft or personnel concerned. . . . Your reports appear to indicate that aircraft rescue is unwarrantedly hazardous, and the force commander concurs in your decision in the matter, pending further developments."

But Balchen and Wimsatt also knew how to play the military's bureaucratic power game. All six men on the ice were army officers and airmen, so they appealed to General Jacob Devers, commander of the Sixth Army Group in Europe. Devers threw his considerable weight behind Balchen's plan, and pressure soon came down on Smith and the navy via the War Department in Washington. With the muscle of the Sixth Army behind him, Balchen outflanked Smith.

Within days, the navy's commander in chief began sending defensive messages such as this one, on January 4, 1943: "[At] no time has it been the intention of Navy Department to withhold use of any Navy facility . . . in undertaking rescue [of] crew of [B-17 PN9E] now down on Ice Cap. Use of Navy PBY airplane under direct supervision of Colonel Bernt Balchen is authorized at any time such action is in accordance with best judgment" of Admiral Smith.

Smith had been boxed into a corner. He recognized that he was being portrayed as hindering the rescue, so he reversed course. He gave Balchen two PBY Catalinas and placed him in charge of the rescue mission. The only conditions were that Balchen had to keep

Smith in the loop and use all-volunteer crews with as few men as possible.

Finding crewmen wasn't a problem, as every man assigned to the PBYs stepped forward. They went to work stripping the armor plating and unbolting the machine guns from the chosen planes to make them as light as possible for the unconventional takeoffs being planned. Then they waited for good weather.

In the meantime, rescuers made one more attempt to use a ski-plane. The results would have been comical if men's lives hadn't been at stake. One plane, a twin-engine Beechcraft, arrived at Bluie East Two on January 20 to have its skis installed. On a trial run, the skis turned upward and were chopped off by the propellers. That ended that.

WITHOUT WORKING RADIOS or walkie-talkies, the only contact between the stranded men and the outside world came during overflights by Pappy Turner's B-17. As the long Greenland winter gathered force, neither the men on the ground nor the crew in the air could predict how many days or even weeks would pass between flights. Turner established a policy of only flying on days with enough light for him to see his plane's shadow on the ice cap. Otherwise, he'd have no idea how high or low he was flying, and his B-17 might end up alongside the PN9E or in Koge Bay.

At the unfinished, undermanned Bluie East Two base, Turner's flight crew had to maintain their own bomber, heating the engine in the frigid predawn hours and improvising when broken parts couldn't be replaced. They had no hangar, so they kept the bomber on the unfinished runway, tying its wings and tail to five-ton trucks to keep the plane from blowing over in gale-force winds. When the starter on the number-two engine failed, flight engineers Carl Brehme and Norman Anderson treated the Flying Fortress like a cross between a child's spinning top and a crank-started Model T Ford. First they attached a rope to the twelve-foot propeller. Then the two sergeants and other crew members set off at a run to pull the rope, spin the blade, and start the engine. Every man among them knew that no replacement planes would be sent to Bluie East Two if they failed. If their B-17 died, so would the men on the ice.

On days when they could fly, they delivered supplies like bombardiers on combat missions. Turner's two flight engineers and radio operator Ralph Coleman hung on for dear life at the open bomb bay doors to push out the packages, one per run. Each drop was free fall, so they'd wrap canisters filled with twenty-five cents' worth of kerosene in expensive padded parkas to prevent them from breaking on impact. In the tail gunner's position, navigator Herbert Kurz fought airsickness and freezing winds to watch where each package landed so the crew could correct for later runs. Kurz also noted which supplies landed too far away, so they'd know which drops to repeat. Based on Kurz's advice, Pappy Turner and copilot Bruno Garr adjusted their routes and brought their bomber down to less than a hundred feet off the ice cap for pinpoint deliveries. This was especially important at what remained of the PN9E, where a cargo drop beyond a few dozen yards from the plane's nose might lure Monteverde, Best, and Spina into crevasse territory.

The flying was routinely treacherous. Although Turner wouldn't fly in storms, he couldn't escape the winds that toyed with the bomber on every run. Adding to the danger, windblown snow racing across the ice cap looked like the top of clouds. Once when he was blown off course and briefly lost his bearings, Turner nearly flew straight into the ground, all the while thinking that he was heading toward a cloud.

In all, Turner's crew made thirty-four supply trips and dropped 225 packages to the snow cave at the PN9E, the Motorsled Camp, and the men and dogs at Beach Head Station and Ice Cap

Station. The supplies could have stocked a country store, including rations; fuel; toothbrushes and toothpaste; medical supplies; fur-lined mitts; woolen underwear; seventeen-inch wool socks; snowshoes; shovels; rope; pup tents; boxes of chocolate bars; grappling hooks; two dozen white handkerchiefs; sleeping bags; soap and towels; boxes of cigarettes; a dozen bottles of Coca-Cola; copper rivets; bundles of magazines; and two picnic hams.

On nearly every trip, in at least one package Turner and his crew tucked a note, either typewritten or scrawled on paper torn from a notebook. Some notes described the latest rescue attempt, some gave bits of news about the war, and some were written just to lift the spirits of exhausted men living like hibernating polar bears. The notes left out bad news, as they were written mainly to boost morale. For instance, nestled under the bomber wing, Armand Monteverde, Paul Spina, and Clint Best weren't told that Clarence Wedel had disappeared in a crevasse or that their three remaining companions were stuck six miles away in the Motorsled Camp.

CAPTAIN KENNETH "PAPPY" TURNER. *(U.S. COAST GUARD PHOTOGRAPH.)*

On Christmas Day 1942, the trio at the PN9E expected Turner's crew to drop them something special, but the supply plane didn't come. They had one remaining can of chicken, but that was a delicacy they were saving to celebrate their rescue. Eating it for Christmas would have felt like giving up. So they had a Christmas dinner of meat from C rations. They sang carols and reminisced about better Christmases they'd known.

The following day, Turner dropped a typewritten note, addressed simply, "B-17." It included the Morse code alphabet, in the event that the men could get their radio working, and also the correct frequency on which they should broadcast. The note read:

> We will keep you well supplied with food, etc., until you are removed to the Ice Cap Weather Station. Both the Motor Sled and the Dog Team are encountering difficulties, which is the reason for the delay. We're sorry that we couldn't be of more aid on Monday [for Christmas], but the high wind velocity made operations almost impossible. Every day that the weather permits we will be in the area dropping supplies and aiding ground operations. . . . Keep your spirits up as we're trying to get you off as soon as possible. Capt. Turner.

THE SUPPLY DROPS at the Motorsled Camp were so effective, and opportunities for exercise so scarce, that Spencer packed thirty pounds onto his trim frame. Tetley added fifteen. Most days, they rose in the dark around seven in the morning. After cooking breakfast, they worked to enlarge and maintain their under-snow quarters. They braved the blustery winds to collect newly delivered packages during the few hours of daylight. They also tried to send messages to Turner's crew by arranging large and small objects in Morse code patterns, but it didn't work. Neither did their attempts to use black oil to spell out words in the snow. They'd cook dinner in the late afternoon, sleep, then make coffee around midnight. They smoked cigarettes when they had them. One day bled into the next.

All the while, O'Hara was failing, shedding pounds faster than his companions were gaining them. He shriveled from a robust young man into a sunken-cheeked wraith. He dipped into and out of delirium. His legs were useless and his body's systems were shutting down. O'Hara couldn't hold down most rations, so Spencer and Tetley fed him soup from cans dropped by the B-17. When those ran out, they boiled rations into a soupy mixture, but that upset his stomach. O'Hara fared better on lucky days when Turner's crew dropped them sandwiches with fresh meat.

Tetley and Spencer were doing their best to keep him alive, but nothing could save O'Hara's frozen feet. Not massages, not fresh dressings, not prayer. Deprived of blood flow, they'd been dying for weeks. Dry gangrene did its witch doctor's work, leaving the twenty-four-year-old lieutenant with blackened, mummified lumps below his ankles. As O'Hara lay helpless in the ice hole, nursed and protected by his friends, living tissue separated from dead flesh.

In layman's terms, Bill O'Hara's feet fell off into his boots.

The last thing he ever felt on his feet were Clarence Wedel's hands sliding across them as Wedel fell into the crevasse.

ON JANUARY 21, 1943, Pappy Turner's crew dropped a note to the Motorsled Camp with a bundle of wooden stakes and fifty red bandanna handkerchiefs. It described Balchen's rescue plan and said a PBY Catalina/Dumbo would try to land on the first clear day. The note instructed Spencer to put his pilot training to work and choose a smooth, crevasse-free landing area and mark it with the stakes. Spencer was told to tie two bandannas to each stake, one at the top and one at the bottom, to create a gauge the PBY pilot could use to estimate distance from the ground. The note ended, "Be seeing you soon." It was signed, "The Boys." Spencer tucked it away for safekeeping.

On January 25, the two PBYs flew from Balchen's base at Bluie West Eight across the ice cap to Bluie East Two. There, Pappy Turner and his crew briefed the rescue fliers about conditions. In the lead PBY, Balchen would act as supervisor and adviser to the pilot, Bernard "Barney" Dunlop, a thirty-one-year-old U.S. Navy lieutenant from Long Island, New York. The copilot was Lieutenant Junior Grade Nathan Waters, with two enlisted men, flight engineer Alex Sabo and radioman Harold Larsen, as the crew. Also assigned to Dunlop's plane was Captain P. W. Sweetzer, the doctor at Bluie East Two, ready to treat O'Hara even before he left the ice. The second PBY would serve as a backup, in case Dunlop's plane went down.

Also aboard Dunlop's PBY was an experienced U.S. Army Air Forces dogsled rescue team consisting of Captain Harold Strong, who'd served in Alaska; sergeants Joseph Healey and Hendrik "Dutch" Dolleman, who'd made their reputations in Antarctica; and nine sled dogs. Balchen respected Strong as an Arctic veteran, and he knew he could rely on Healey and Dolleman, who'd been part of the trail team in the *My Gal Sal* rescue.

The crews and the planes were ready, but the weather wouldn't cooperate. When January ended with no relief at the Motorsled Camp, Don Tetley grew depressed. A mission that he thought might take a few days had stretched to two months, with no end in sight. Two men had fallen to their deaths in crevasses before his eyes. As days turned to weeks, he focused his remaining hopes on being rescued before February 1, the birthday he shared with his wife.

Tetley's target date arrived with no sign of a PBY. He became morose, thinking about what he was missing back home and how worried his wife must be. The Motorsled Camp was out of cigarettes, so Tetley glumly passed the hours collecting tobacco from spent butts. Sizing up his partner's mood, Harry Spencer sprang a birthday surprise. When they'd been flush with cigarettes, he'd set

aside a full pack for just such an emergency. Now he made a fuss
of presenting it to Tetley with birthday wishes.

The gift was one of untold acts of kindness, large and small,
that the men bestowed upon each other. Each one revealed a
bond that was crucial to their survival. With every sacrifice, every
shared cigarette, the icebound men expressed a stubborn refusal to
surrender their humanity.

LATER THAT DAY, Turner's B-17 crew dropped Spencer the best
possible reward for his thoughtfulness short of rescue: a pack of
mail, including letters from his wife, Patsy, and a sweater she'd
knitted him for Christmas. With it came another note from "The
Boys," saying that the PBYs had reached Bluie East Two and
were waiting to fly. "We will try to get you out this time," the
note read. Tetley saved that one.

Two days later, the supply drop included a new walkie-talkie
freshly arrived from the United States. The "two-way air-to-
ground communication system," as it was known, was in high demand, especially in battle zones. Now that the Motorsled Camp
had one properly tuned to Pappy Turner's B-17, the airborne cargo
team and the grounded men could finally talk to one another. At
first, the men's cold breath froze the mouthpiece, so they thawed it
out and put a sock over it. Then they introduced themselves to the
air crew that had kept them alive. The contact came as a relief to
Pappy Turner, who'd felt frustrated that he could "be so near those
men that you could nearly touch them, and yet you couldn't do
anything about getting them out."

Yet the wait continued for Balchen, Dunlop, and the PBYs.
The winds were too treacherous that day and also the next. Blinding sheets of snow raced across Spencer's chosen landing field, the
drifts obscuring the bandanna-topped stakes. Turner and Spencer
talked over the walkie-talkie about surface conditions, and Turner
asked whether his crew should drop more spare parts for the mo-

torsled. Spencer and Tetley had last worked on the machine more than a month earlier. Since then, it had disappeared under mounds of snow.

"For crying out loud," Spencer answered, "we've been digging for two weeks and can't find a trace of it!"

O'Hara's hopes had risen when he'd learned of the latest rescue plans, but now he felt crushed by the delays. Emaciated and dehydrated, his skin a waxy yellow, his feet gone, he grew depressed. Spencer and Tetley feared that O'Hara might give up and die before the PBY arrived.

18

SHITBAGS

AUGUST 2012

"WE NEED SHITBAGS."

"Sorry, Lou, I didn't hear you."

"Shitbags. We need shitbags. Get as many as you can."

With that phone call, Lou sends me on a final errand before I join him and the North South Polar team for the start of the Duck Hunt. To satisfy his request, I visit a camping supply store to buy a glacier-load of disposable bags to fit a toilet that, minus bags, is just a plastic seat with a hole in it.

With seven dozen shitbags, a duffel bag filled with cold-weather camping gear, and a bottle of Shackleton's reconstituted Scotch, I drive five hours from Boston to Somerset, New Jersey. There, Lou's second-in-command, a logistics specialist and retired Army Reserve colonel named Steve Katz, has surrendered his suburban home for use as a staging base, to make sure that everything we need is in working order before we reach the ice. We're leaving in two days aboard the C-130 from the Trenton-Mercer Airport, not far from Katz's house, bound for Keflavík, Iceland, en route to Kulusuk, Greenland, en route to a glacier at Koge Bay, in search of the Duck's resting place.

Neighbors drive wide-eyed past Katz's tidy blue house, not sure if he's holding a massive yard sale, planning a military assault, or hosting a traveling circus. The truth is a little of all three. Dominating Katz's front lawn is an orange-and-gray dome tent called a Space Station, more than nine feet high and twenty feet in diameter. A yellow canopy stretches from the dome to a rectangular green tent bursting with camping and survival gear in various states of assembly. Across the lawn, a tall, slender young man—Lou's twenty-two-year-old son, Ryan Sapienza—is erecting a gray tent the size and shape of a Porta Potti. Next to Ryan is a green plastic folding stool with a hole where the seat belongs. The shitbags have found their mate.

Lining both sides of the driveway and overflowing onto the lawn are twenty-five black plastic Pelican cases, each four feet long and two feet high. It looks like an outdoor showroom of open-lid caskets for rotund short people. Inside the big Pelicans and numerous smaller cases are water and fuel tanks, hoses, lights, ice axes, generators, coats, boots, sleeping bags, sleeping pads, gloves, boot dryers, doormats, freeze-dried food cans, ice drills, satellite phones, climbing gear, cooking supplies, plates, cups, metal utensils, clothespins, a rifle, aluminum foil, hand sanitizer, fire extinguishers, canteens, lighters, North South Polar baseball caps, baby wipes, glacier glasses, batteries, coffee makers, sunscreen, socks, shotgun shells, peanut butter, mission patches, paper towels, a shotgun, playing cards, Clorox bleach, spatulas, stoves, cookie sheets, toilet paper, Gatorade, ziplock bags, flashlights, waterproof pants, polypropylene underwear, and fleece jackets. And that's not even the half of it. Nearly everything in sight has been tagged with yellow stick-on labels that identify the intended user or the object itself. Some wise guy has pasted a label on the handle of a pickax that reads, "PICK AX."

Despite the mountain of gear in Steve Katz's front yard, the shopping spree continues; it turns out that shitbags were hardly

the last missing items. Team members disappear to buy more supplies, from reams of paper to rulers, fresh food to more clothing.
Trucks from the U.S. Postal Service, UPS, and FedEx crowd the
winding street, disgorging boxes in all sizes. By the time we're
ready to leave, more than eight thousand pounds of gear will ac-
company us to Greenland, or about five hundred pounds per expedition member. Two thoughts cross my mind: the men in the
PN9E survived for months with almost none of this, and now I
know where my money went.

Playing joyful ringmaster is Lou, moving from one cluster of
supplies to the next, holding a cell phone to his ear as he sets up
a portable expanding flagpole. On it he hoists three banners: the
American flag, the black-and-white POW/MIA flag, and a red
flag he had made with a gold star and the words "Honor and Remember" above the names Pritchard, Bottoms, and Howarth. Lou

LOU SAPIENZA CHECKS ON EQUIPMENT DURING PREEXPEDITION PLANNING AT STEVE
KATZ'S HOUSE. *(MITCHELL ZUCKOFF PHOTOGRAPH.)*

turned sixty a week ago, and he's balanced on knees that need replacing. But right now he's like a kid at Christmas.

He booms a welcome and throws an arm around my shoulder. "Let's take a walk," he says. Not five minutes after my arrival, Lou asks if I'd consider using my home as collateral for a bridge loan until the money from the Coast Guard arrives at the end of the expedition. He's tapped out, and he needs to pay vendors for the supplies still arriving. I swallow hard and agree. In the end, the loan doesn't come through, so instead we rely even more heavily on my credit card.

Moving with ant-farm enterprise around the yard and inside Katz's home is the North South Polar crew. "They're like the team from *Armageddon*," Lou says, referring to the movie in which Bruce Willis saves the world from an asteroid with a gang of rough-and-ready eccentrics.

First among equals is Robert "WeeGee" Smith, a mechanical wizard who builds rally cars in Colchester, Vermont. The nickname WeeGee dates from his childhood, when an older brother couldn't pronounce his prior nickname, Luigi, bestowed on young Robert in tribute to his Mediterranean complexion. WeeGee and Lou worked together in 1990 and 1992 during the Greenland excavation of the Lost Squadron P-38 known as *Glacier Girl*. For months during that project, WeeGee spent most days 268 feet below the glacier surface, carving the plane from its icy tomb: "I had never before worked in a place where it can kill you in a second, without batting an eye. It was great." Long divorced, with an adult daughter and a teenage granddaughter, WeeGee is fifty-nine. His trim build, limitless energy, unlined face, and bright green eyes make him seem twenty years younger. He has a reputation for outhustling everyone around him, for refusing to suffer fools, and for having an almost mystical ability with machines. WeeGee's primary job on the ice will be to operate an industrial-sized hot-water pressure washer called a Hotsy to melt holes deep into the glacier to investigate radar anomalies that might be the Duck.

Working nearby is Jaana Gustafsson, a forty-three-year-old Finn who lives in Stockholm with her husband and two daughters. Tall and attractive, with an engaging smile and a PhD in geophysics, Jaana (pronounced *Yah'*-nuh) is new to North South Polar. She's already earned WeeGee's respect by helping him hoist loaded Pelican cases weighing up to three hundred pounds. When I admire her fortitude, Jaana teaches me the Finnish word *sisu*, which translates roughly as perseverance but speaks more to strength of character. A land surveyor by profession, Jaana is an expert on ground-penetrating radar. On the ice she'll strap herself to a radar unit made by her former employer, MALÅ Geoscience, with a flexible thirty-five-foot-long "dragon tail" antenna that she'll drag across the glacier. Jaana's radar work is supposed to confirm or rule out the anomalies identified by airborne surveys. One major problem: the radar equipment, shipped via UPS from Sweden, is hung up in customs.

Also in Steve Katz's driveway is W. R. "Bil" Thuma, at sixtynine the oldest member of the team, an endearing curmudgeon with a white Brillo mustache and a round belly he displays by opening his shirt in the August heat. An American-born citizen of Canada, Bil is a former *Glacier Girl* team member with fifty years of geophysical fieldwork and longtime expertise in the under-ice landscape of Greenland. Bil earns his living as a consultant, marketing technology for natural resource exploration in places like Libya, Kazakhstan, and Mongolia, his destination following the Duck Hunt. At the edge of the driveway, Bil furrows his brow as he reviews satellite photos of the Koge Bay glacier. He's worried that the expedition will be a bust before it begins. "This whole area looks heavily crevassed," he says. "I don't know how Jaana can run the radar—if it arrives—without falling into one."

Keeping Jaana and the rest of us out of crevasses is the job of

North South Polar's safety team, led by Frank Marley, the just-back-from-Afghanistan Army National Guard captain who earlier strategized by e-mail about polar bear defense. Frank is forty, easygoing, a solid mass of muscle and an expert outdoorsman. He's also a third-year medical student with plans to specialize in expedition medicine. In Greenland, one gun will always be near Frank, just in case. Fellow safety team leader John Bradley, with close-cropped reddish-brown hair and the start of an expedition beard, is a mountain rescue expert, a veteran of lifesaving missions on Denali in Alaska, Mount Whitney in California, and Pico de Orizaba in Mexico. For a day job, John heads the climbing department at the Denver flagship store of outdoor retailer REI. Working alongside Frank and John is Nick Bratton, a veteran ice climber and mountaineering guide from Seattle whose day job involves designing land conservation programs. Tall and lean, with long strides balanced on size-fifteen feet, Nick's peripatetic past includes a year working as a whitewater rafting guide on South Africa's Tugela River, which he captured in a book called *Guided Currents*. Nick, who's married to a mental health counselor, soon reveals himself to be the most safety-conscious person I've ever met; he'll wait five minutes at the corner of a deserted intersection until the walk signal turns from red to green.

Lou's son Ryan was supposed to join the safety team, but an accident this summer at a restaurant where he works almost severed two fingers. He'll help manage the base camp and maintain the expedition's daily log. An adjunct member of the safety team is Michelle Brinsko, a thirty-eight-year-old blond, blue-eyed physical therapist from Ohio with a sweet nature and fearsome biceps. Michelle is the Duck Hunt's cook and provisions chief—she carries a notebook of camp recipes with the mission statement: "Failure is not an option"—and is in a relationship with Frank Marley.

After me, next to arrive is Steve Katz, just back from driving his older daughter to college in Virginia. Nearing fifty, with dark, thinning hair, Steve is built like a thick-shouldered fireplug. As a colonel in the Army Reserves, he led a Special Forces unit during the Iraq War and earned two Bronze Stars. Soon after Steve's arrival, I learn that he's also helped to cover expedition costs with a credit card.

The final member of the North South Polar team will join us before takeoff: Alberto Behar, the expedition's chief scientist, a man whose résumé can make almost anyone feel inadequate: PhD in electrical engineering; two-decade career at NASA's Jet Propulsion Laboratory, including oversight of an experiment on the Mars rover *Curiosity*; robotics expert; rescue scuba diver; helicopter and fixed-wing pilot; emergency medical technician; faculty member at Arizona State University. Forty-five years old, married with three children, Alberto has been designated "highly qualified" to become a NASA astronaut and is awaiting word on an interview. His curly black hair and handsome face prompt Lou to call him "rock star." Alberto has built a high-definition video camera that can be dropped into the holes that WeeGee melts in the ice, to determine whether an anomaly is the Duck or a hidden crevasse, a pool of water, or an otherwise false reading.

WHEN WE REACH the glacier, Lou's plan calls for Jaana and Bil to use ground radar to confirm or rule out the anomalies identified by the air surveys; WeeGee to melt holes to any sites that look promising; Alberto to drop his camera down the holes for a look; Frank, Nick, and John to keep us safe; Michelle to keep us well fed and hydrated; Ryan to record everything that goes on and help manage base camp; and Lou and Steve to supervise, with me pitching in wherever needed. All this is supposed to happen during approximately one week at Koge Bay, regardless of foul weather, rough terrain, or technical problems.

After Lou explains what's planned, private conversations reveal the team's shared anxiety about the long odds against finding the

Duck. Everyone here knows that the 2010 mission was a bust, and no one, except perhaps Lou, expects the Coast Guard or anyone else to fund another attempt if this trip fails as well. After innumerable "needle in a haystack" descriptions of what we have planned, a new metaphor evolves: imagine searching for a diamond chip buried deep beneath a frozen football field; your best tool is a straw that makes tiny holes into the ground, through which you peer down to see what's below; if your holes miss by even a little, you'll never find it; and you have a brief window to explore ten potential locations before being kicked off the field.

Lou says it's a sure thing; everyone else has doubts.

At the moment, though, nothing will happen beyond our suburban staging area if we don't get the radar equipment. Lou yells into his cellphone at a representative for UPS, which is supposed to get the radar through customs for delivery to us: "This is a Coast Guard operation. We've got an $8 million plane waiting for us, and you're holding it up!" The UPS agent is apologetic, but can't say when the radar might be released. Steve decides to remain behind when the C-130 leaves; he'll carry the radar equipment on commercial flights when it clears customs.

THE FINAL TWO days before departure are a whirlwind of last-minute purchases, loading, and preparations, with team members renewing friendships or getting to know one another. The day before the C-130 arrives, WeeGee rents a truck to transfer our gear from Steve's house to Trenton-Mercer Airport.

As WeeGee and I drive to the rental lot, we exchange life stories and WeeGee asks about the book I'm writing. When I explain that I'm telling the historic story and also describing this expedition, a sly smile crosses his face.

"So Mitch," he asks, "how does it end?"

"No idea, WeeGee. You tell me."

The exchange becomes a routine between us, sometimes spo-

ken several times a day, particularly when our prospects seem the
most bleak.

EARLY ON AUGUST 21, 2012, one day later than planned, we
drive in convoy to the Ronson Aviation hangar at the Trenton-
Mercer Airport. I'm in front with Bil, Jaana, and Nick in my
car; WeeGee and John drive the truck; Frank and Michelle ride
in Frank's car; and Lou and Ryan bring up the rear. Steve is off
somewhere trying to find the radar, and Alberto is meeting us at
the airport. I take a wrong turn out of the hotel parking lot and
get razzed by my teammates, a bad omen.

A half hour after we arrive at the airport, I see a vision in a lilac
pantsuit: Nancy Pritchard Morgan Krause, the sister of pilot John
Pritchard, here with her husband, Bill, to see us off. Lou invited
her several days ago, so she and Bill drove up from their home in
Annapolis.

As Nancy hugs me hello, I notice a bruise on her forehead and
a bandage on her arm. She whispers that she fell a day earlier when
checking into their hotel; later she admits that she passed out,
possibly from the heat, and hit her head. "They wanted me to stay
in the hospital," she says. "But I wouldn't. I wouldn't miss this."

Her eyes grow misty when Lou shows her the flag with her
brother's name on it. Bill stands close by her side to keep her
steady. One after another, North South Polar team members in-
troduce themselves and have their photos taken with her. Nancy
quizzes each one about his or her role, then thanks everyone for
trying to find her big brother.

"He loved what he was doing," she says, "and he was an outgo-
ing, friendly person. He would do anything to help people." Several times Nancy repeats the story of how devastated she felt when
she learned that John was lost. Afterward, Nick Bratton writes in
his journal, "We are contractors to the Coast Guard, but our real
client is Nancy."

Nancy and Bill are driven by golf cart onto the runway, where the Coast Guard C-130 is waiting for its crew to load our gear. Nearly one hundred feet long, with four propeller engines hanging beneath its wings and a yawning cargo bay open beneath its tail, the gleaming white-and-orange plane is our ticket to Greenland.

Along with its crew, the C-130 disgorges the Coast Guard team joining us on the ice. All five members seem like natural additions to Lou's real-life cast of a Bruce-Willis-saves-the-world movie. The leader is Commander Jim Blow, fulfilling his role as the service's point man on the mission. Next is Lieutenant Commander Rob Tucker, tall and good-humored, a pilot who works with Blow in the Office of Aviation Forces. Documenting the mission will be Petty Officer Second Class Jetta Disco, a cheerful bundle of energy from the service's New York public affairs office. Our medical officer will be Captain Kenneth "Doc" Harman, a flight surgeon, an-

BEFORE THE C-130 LEAVES FOR GREENLAND, (FROM LEFT) W. R. "BIL" THUMA, LOU SAPIENZA, JIM BLOW, AND NANCY PRITCHARD MORGAN KRAUSE REVIEW THE MISSION PLAN AND THE MAPS OF KOGE BAY. *(U.S. COAST GUARD PHOTOGRAPH BY JETTA DISCO.)*

tique boat restorer, and raconteur with experience in trouble spots around the world. The Coast Guard also has brought along Terri Lisman, a geophysicist from the National Geospatial-Intelligence Agency. An accomplished swing dancer and home-brewing enthusiast, Terri will be the government's own radar expert on the ice.

When Nancy sees Jim Blow, she wraps him in a hug and describes what the Duck Hunt means to her: "You think of the plane and the bodies there, alone. If you bring them back, they're home. It's closure." Lou strolls over, and he and Jim outline the mission plan, using an oversize laminated map of the Koge Bay glacier to show Nancy where we hope to search.

As the heat rises and Nancy tires, Lou organizes an impromptu ceremony in the shadow of the C-130. Nancy's husband, Bill, presents Lou with a gift: "I'm sure everything is going to be a success, but I brought you a tool to pull it together if things threaten to fall apart." He hands Lou a rubber band, and everyone laughs.

Nancy bestows a final blessing on the now-complete Duck Hunt team, a dozen of us with North South Polar and five with the Coast Guard: "God go with you, and bring you home safely and successfully."

As Nancy and Bill prepare to leave, Steve arrives, having somehow pried Jaana's ground-penetrating radar equipment away from UPS and the U.S. Customs Service. We find seats in the C-130, with our Pelican cases, personal gear, and other supplies strapped in the cargo bay behind us. A Coast Guard crewman hands out earplugs for the seven-hour trip, and the Duck Hunt takes flight.

19

DUMBO ON ICE

JANUARY–FEBRUARY 1943

THE NEW YEAR started as poorly as the old year ended for the three men in the snow cave under the wing at the PN9E. Unrelenting storms spat snowflakes as sharp as needles and kept them hunkered deep in their sleeping bags. On what they calculated was January 2, pilot Armand Monteverde awoke in the dark feeling heavy pressure against his body. He struck a match and saw snow pouring in through a small hole in the windward wall. Enough snow had already accumulated to make him fear being buried alive.

A wind-whipped blizzard carved more holes in the wall, through which fine, stinging flakes rushed in. The winds came from the north, building speed and strength across hundreds of miles of featureless ice. Monteverde woke Clint Best and Paul Spina, and together they used what they had at hand to plug the holes. But it was like fighting the tide with a bucket. The snowdrifts grew higher around them, and the constant battering threatened to collapse the north wall of their shelter.

For the first time in the two months since the crash, Monteverde, Best, and Spina abandoned hope. They'd been down before, but never as far down as this. They knew that winter storms in Greenland could blow for days without pause. If this storm was one of those, there was no point spending their last precious energy on a fight they couldn't win. They shook hands and prayed together, then prepared for the end.

The trio moved into the entrance tunnel in the cave's south wall and spread out their sleeping bags. Exhausted and defeated, Best climbed into his bag and fell fast asleep. Monteverde's sleeping bag was half frozen, but he got in anyway. Numb with resignation that he was about to die, the shivering pilot closed his eyes and surrendered to fate. Spina lay awake. For hours he replayed his life in his mind's eye, from his childhood in rural New York to what he expected would be his miserable end as a human icicle in Greenland.

Time slipped by in the dark. The wind snuck under the plane's wing and lifted it from the ice wall, opening the PN9E shelter to the elements and allowing snow to pour in. Spina woke his companions, but they agreed there was nothing to do but pray. After a while, they noticed that less snow was accumulating in one corner, so they moved their sleeping bags there and tried to sleep.

Hours passed and all became quiet. The storm blew itself out without burying them alive.

The three men awoke, surprised to find deep snow covering everything except the corner where they'd crammed together. They made coffee and went to work rebuilding and reinforcing the north wall. They melted snow in milk cans and poured the water on the rebuilt wall, so it would freeze solid. They reinforced it with boards from supply crates. When they were finished, the wall was stronger and thicker than ever, and even the fiercest storms wouldn't penetrate. They left several small holes at the top of the wall, to let in light during the day. At night, they filled those "windows" with plugs made from rags and trash, to block the wind. At Spina's suggestion, they left much of the accumulated snow inside their cave, so they wouldn't have to go outside all winter to gather it for cooking and drinking.

As the days dragged by, the glacier on which the trio lived continued its crawl toward the sea, spawning new crevasses as its forward edge sought rebirth as icebergs. Storms and windblown snow built fresh bridges over the deep gashes, so each time the men went outside to fetch new supplies, they knew that any step could be their last. Their rebuilt quarters were only marginally safer. The PN9E's tail section had already been swallowed, and nothing could prevent a new crevasse from opening beneath them. All three felt tremors as the ground shifted, but they didn't talk about it. There was no point.

ON FEBRUARY 5, 1943, the weather on the east coast of Greenland was as good as it gets in winter. By Harry Spencer's estimate, it had snowed a whopping eighteen feet at the Motorsled Camp since they'd been there. But this was a rare day with clear skies. A scout plane flew overhead and reported to Colonel Balchen that at eight o'clock in the morning the winds were calm, the sky was blue, and the ground temperature was about 10 degrees below zero. Greenland's version of balmy.

Balchen told his rescue teams to get ready. But just as he ordered the PBY Catalina crews to climb aboard their planes, a messenger handed him a troubling radio dispatch from General Henry Harley "Hap" Arnold, commanding general of the U.S. Army Air Forces. It read, "Factory indicates forward bulkhead of PBY too weak for landing on snow." The forward bulkhead, Balchen knew, was a wall near the nose of the plane. If it buckled under pressure from a belly-down landing, the PBY would suffer a structural collapse, destroying it and possibly the men and dogs inside. Arnold's note had one question: "What are you[r] plans?"

Balchen scribbled a reply: "Going ahead as contemplated." He explained later: "We have had no time to make a test landing, but I figure that if anything is going to happen it will happen anyway, test landing or not."

THREE PLANES TOOK off that morning from Bluie East Two. Pappy Turner's B-17 led the way and the two PBYs followed close behind. Because they were on a rescue mission, on this day the PBYs would properly be called Dumbos. When they were over the Motorsled Camp, Turner radioed down to Harry Spencer on the walkie-talkie to discuss ground conditions. Turner worried that the winds were too strong for a PBY to land, but Spencer assured him that it was safe.

At the controls of the lead PBY, Navy Lieutenant Barney Dunlop took one long run over the landing area that Spencer had marked with stakes. On the ground, Spencer and Tetley watched from outside their ice cave as the plane's nose seemed to dip. Unwashed and unshaven, their heavy beards coated with snow, they looked like a pair of prehistoric icemen hunting a flying dinosaur. When he saw the plane's nose drop, Tetley feared that the PBY was in peril. But Spencer realized that the plane was hidden behind a small rise and was coming in for a landing.

Earlier, when reviewing the plan with Dunlop, Balchen had explained that the landing area was mostly flat, so he suggested that Dunlop bring the plane down at normal speed, "like a power stall letdown on a glassy sea." The technique, Balchen hoped, would keep the PBY's nose high, preventing it from burrowing into the snow. With a gradual rate of descent and a steady hand on the control wheel, Balchen believed, the plane would almost land itself.

Now, just as they'd planned, Dunlop flew at an air speed of about eighty knots, or ninety-two miles per hour. As Turner's B-17 and the backup PBY circled overhead, Dunlop brought his Dumbo down with the wheels retracted and the wing floats down. The twenty-odd seconds it would take to go from the sky to the ice cap would be just long enough for Balchen to recall the PBY manufacturer's warning about potentially catastrophic weakness in the forward bulkhead.

The plane's altitude above the glacier dropped to zero, and the PBY's rounded hull grazed the snow. Dunlop cut the throttles. The propellers sprayed sparkling clouds of frosty mist. The full weight of the plane settled onto Greenland, and the bulkhead held strong. Dumbo was on ice.

Balchen's plan had worked, at least so far, and Dunlop's flying skills had proved first-rate. By the time the plane stopped, its keel had carved an eighteen-inch-deep scar across Greenland's face.

As he watched, wrung out from all he'd been through, Harry Spencer experienced a strange absence of feeling. Happiness about the Dumbo's arrival mixed with sadness for all the men lost, and the two emotions canceled each other out. This was the eighty-eighth day since his bomber had crashed. In that time, two crew-mates, Loren Howarth and Clarence Wedel, had been killed, along with three men who'd set out to rescue them: John Pritchard, Benjamin Bottoms, and Max Demorest. He knew that the five men in the C-53 that he'd hoped to find remained lost and were presumed dead. Spencer himself had survived a hundred-foot fall into a crevasse, and he'd watched helplessly as O'Hara had lost his feet. Now, the sight of the PBY pulling to a stop to carry him home-ward was almost too much to process.

Tetley's emotions were less conflicted. He felt as though he'd willed the plane to a safe landing. Standing alongside Spencer, he felt soothed by what he called "a beautiful sight."

The crew poured out of the plane and hustled to the hole to help O'Hara. With Dr. Sweetzer by his side, Balchen described finding the young navigator in his sleeping bag, able to manage a wan smile. O'Hara had lost half his normal body weight of 180 pounds, leaving him as "light as a bundle of rags," Balchen said. Balchen would recall carrying O'Hara aboard the plane, but Spencer said they used a specially built stretcher-sled. They were too busy for much chatter, but one piece of information the rescuers shared with Spencer and Tetley was that the three men left behind at the PN9E were apparently still alive.

ABOUT FIFTEEN MINUTES elapsed between the landing and Dunlop's return to the cockpit to prepare for takeoff. When everyone was in place, he leaned on the throttles. The engines revved and the propellers spun, but the plane wouldn't move. Dunlop tried again, with the same result. In the brief time since landing, the Dumbo had frozen to the ice. Greenland wasn't done with them yet.

Balchen ordered the able-bodied men off the plane and onto the ends of the wings. They jumped up and down, and after a while they broke the ice and freed the plane. But it froze again to

BARNEY DUNLOP'S PBY DUMBO AFTER A BELLY-DOWN LANDING ON THE ICE CAP. *(U.S. ARMY PHOTOGRAPH BY BERNT BALCHEN.)*

the glacier before they could climb back inside. On a second try, Balchen positioned the men on the ground at the PBY's two wing floats. On his signal, they rocked the plane like a seesaw while Dunlop advanced the throttles. After almost two hours of effort, the glacier released its grip.

Dunlop taxied in a wide circle, knowing that if he stopped the plane would freeze again to the surface. The ice-busting crew ran alongside the Dumbo, racing to reach a rounded protrusion on the side of the fuselage aptly called a "blister." The blister had a door in it, and from inside radioman Harold Larsen reached through it. Outside, the men on the ground fought the powerful wash created by the propellers. They ran toward the blister, each one jumping at the last minute with his arms outstretched. Larson caught one after another and pulled them inside like parachutists in reverse.

Dunlop took off without incident, and the brief flight was uneventful. When the PBY landed at Bluie East Two, Sweetzer rushed O'Hara to the medical ward for the first of what would be several long hospital stays. In the months ahead, surgeons at Walter Reed Army Medical Center would complete the job begun by gangrene: they amputated what remained of both of O'Hara's legs below the knees.

Spencer and Tetley arrived at the base in remarkably good shape, suffering mainly from fatigue, exposure, and weight gain. Spencer had become so hefty that he split his pants during the rescue. They spent two days in the medical ward, during which they donned new uniforms, shaved their beards, and prepared to fly home to the United States for long leaves.

With that, Harry Spencer and Bill O'Hara joined Al Tucciarone and Woody Puryear as members of the PN9E crew to escape Greenland's grasp. The tally of the original nine-man B-17 crew was four rescued, two dead, and three still waiting.

AT THE PN9E camp, Monteverde, Spina, and Best were on the verge of giving up again.

The stove Tetley left behind had burned out before the end of January. The last three B-17 crash survivors returned to the more primitive cooking method of lighting the bomber's leftover fuel in the bombsight case. They only had leaded gas, which gave off noxious vapors, so they usually used it outdoors.

Meanwhile, Pappy Turner's crew finally received a walkie-talkie they could drop to the PN9E. Nearby was a handwritten note: "Hello boys, Get on the Walkie-Talkie we're dropping with this right now. Let's talk!" Below was a drawing of the device, with detailed instructions. The note signed off: "We won't quit until you're with us."

The walkie-talkie broke in the fall, but the men fixed it by cannibalizing parts from the useless one that Tetley had left behind. When they connected, Turner's crew told them about the plans

HARRY SPENCER (LEFT) AND DON TETLEY ABOARD THE RESCUE PLANE. *(U.S. ARMY PHOTOGRAPH, COURTESY OF CAROL SUE SPENCER PODRAZA.)*

to use PBY Catalinas in a rescue attempt. In a later conversation, Turner told the men at the PN9E that their crewmates were off the ice.

Soon after the walkie-talkie's arrival, however, the three men at the PN9E had no one to talk to but each other. During most of February, Greenland's weather behaved as though enraged by the audacity of Balchen's plan and Dunlop's flying feats. Blizzards roared to life for three weeks, during which no planes could fly.

With their stove broken, the remaining PN9E survivors had no reliable heat. Soon they ran out of candles, so they spent long stretches in darkness. Desperate for warmth and light, they kept a fire burning in the bombsight case. Without a vent, toxic fumes filled their cave and soot blackened their skin and clothes. With no other way to defrost their rations, Monteverde, Spina, and Best

HARRY SPENCER (LEFT) AND DON TETLEY SHORTLY AFTER THEIR ARRIVAL AT BLUIE EAST TWO. *(U.S. ARMY PHOTOGRAPH, COURTESY OF CAROL SUE SPENCER PODRAZA.)*

burrowed deep into their sleeping bags and held the cans and packages under their armpits. Eight hours of this made the food soft
enough to chew. Some ration cans broke when they were dropped
from Turner's B-17, and when they thawed the juices leaked onto
the men's bedding. The smell of rotten food mixed with the stench
of burned fuel, body odor, and human waste.

Spina's broken right arm continued to pain him, but his frostbitten left hand regained some feeling and movement. However,
all of his fingernails had fallen off, so it hurt whenever he touched
something with the tips of his fingers. Monteverde's feet remained
painful, but he hobbled around as much as he could. Physically,
Best was the most able among them.

They hadn't bathed or shaved since the crash, but Turner
dropped fresh uniforms, so they made a practice of changing their
clothes and underwear at least once every two weeks. They hated
stripping down in the cold, but clean underwear and socks always
made them feel warmer. As soon as they dressed they'd climb back
into their sleeping bags. Spina laid his head on a five-gallon can of
dog food. The sled dogs hadn't reached them, so the can remained
full. Monteverde and Best each used an airman's boot as a pillow.

To fight cabin fever, they played word games. They named all
the countries, rivers, capitals, islands, and every other geographical
feature they could think of. They told and retold their life stories
and talked about whatever came to mind. Still they ran out of
things to say, so they spent long periods in silence. The isolation,
the wind, the moving glacier beneath their cave, and the relentless
cold preyed on their nerves. They seemed to take turns breaking
down, wishing their ordeal were over, one way or another. Each
time, the other two would comfort the crying man. When the
cycle unraveled, all three sank into despair at the same time. They
hatched a suicide pact.

They'd been on the ice cap for almost three months. Three men
who'd been sent to rescue them were dead, as were two of their

crewmates. As Spina put it: "Why should someone else stick his neck out to save ours?" They talked about tying themselves together and hurling themselves into a crevasse. Or maybe they'd leave their igloo and walk until they dropped from exhaustion. Finally, they decided to go quietly. They'd stop fighting, stay in their hole, and let Greenland freeze them to permanent slumber. They agreed that the next time Pappy Turner's B-17 flew overhead, they'd tell him to end his supply flights, cancel all ongoing rescue plans, and, in Spina's words, "scratch our names off the books."

Days passed before the next supply flight arrived, and in that time they reconsidered. They steeled themselves and agreed that suicide was a coward's way out. They vowed not to break down again, but it was a vow they couldn't keep. They resumed the cycle of hopelessness, as one man gave up, then the next, and then the next. Pappy Turner's plane returned as they wallowed in despair. They told him via walkie-talkie that they appreciated all that he and his crew had done, but supply- and morale-sustaining services were no longer required.

Pappy Turner couldn't believe his ears. Normally even-keeled, he flew into a rage. He and his men had busted their humps to keep the PN9E crew alive, and now they wanted to die? Turner bawled them out, telling the trio that until now he'd thought that they had guts. He'd thought that they were strong enough to stick it out, that they were soldiers, but he must've been mistaken. Gaining steam, Turner called them "a bunch of weaklings." He told them he was so revolted that he felt half tempted to accept their plan and let them freeze or starve to death. But the choice wasn't his or even theirs to make. Turner told Monteverde, Best, and Spina that they were the last three pieces of the most expensive rescue of the war. The U.S. government had invested too many lives, too much time and effort, and too much money to let them die now. When he calmed down, Turner promised that he and his crew wouldn't stop flying until the three men were off the

ice. In exchange, he made Monteverde, Best, and Spina promise
that they'd refuse to quit.

Even if Turner was exaggerating for effect, his outburst had
the desired result with Armand Monteverde and Paul Spina. Clint
Best was another story.

SEVERAL DAYS AFTER Turner's tirade, Best sat in the ice cave,
warming C rations over an improvised stove made from a can
filled with leaded gasoline. He usually cooked carefully, knowing that the flames might ignite nearby tanks of gas or even the
fuel-filled wing above them. But Monteverde and Spina watched
as Best sat motionless as food atop the can began to burn. They
called to him, but Best didn't react. As flames rose, Best stared
blankly at his companions. Spina felt chills down his spine. He
called it "the coldest look I ever seen in my life." Monteverde
ignored Best and snuffed out the fire himself.

Best began to shake, so they covered him with blankets. Still
he shook. Monteverde thought Best's strange behavior might be
a delayed result of the head injury he suffered in the crash. Not
knowing what else to do, Monteverde and Spina agreed that a
shock might snap Best from his catatonia. Monteverde snuck up
from behind and slapped Best in the face. Stunned, Best emerged
from his trance and asked what had happened. When Monteverde
explained, Best said he felt as though he'd been lost in another
world. In his hallucination, he told his companions, he was surrounded by people saying that he'd abandoned his post and would
be court-martialed for going AWOL.

As Best told his story, Spina saw the glassy, vacant stare return to his eyes. Best began to shake again. They zipped him
into his sleeping bag and covered him with parkas. When Best
fell asleep, Monteverde and Spina discussed whether to tell
Pappy Turner about Best's breakdown. They decided to wait and
hope that he snapped out of it, knowing that Best had applied

for officer candidate school. Word of mental problems could destroy that dream.

They tried to sleep, but Best and Spina bunked next to each other on the floor. Whenever Best shook, Spina woke. Spina stared into the darkness as Best twisted in his bag. Best's delirium returned and his movements grew erratic. He yelled that three men were fighting with him. Defending himself against phantoms, he reached out and grabbed Spina's left hand. Fearing that Best might break his good arm, Spina called for help. Monteverde slapped Best a second time, and again Best woke from his trance and asked for an explanation. They told him that he'd had a bad dream and that everything was fine. Monteverde and Spina stayed awake, watching Best as he slept, shook, and sweated through the night.

The next morning, Best remained in a trancelike stupor. The other two tried to keep him covered, but he ripped away the sleeping bag and pressed his head against the snow, as though trying to cool his fevered mind. When they offered him breakfast, he accused them of trying to poison him. Best muttered throughout the day, "talking about things drawn from another world," as Spina put it.

That night, Monteverde and Spina alternated keeping watch over Best. During the first shift, Spina heard Best stumbling around in the darkness. Spina turned on a flashlight that they saved for emergencies, but he couldn't find Best in the small cave. He called to Monteverde, who grabbed the flashlight and went to the entrance tunnel. The flap at the far end was open.

Best was gone.

20

ICEHOLES

AUGUST 2012

FIVE HOURS INTO the C-130 flight, we crowd against the windows to watch the sunset over the southern tip of Greenland. Jagged, gray-black mountains rise at the coastline, and beyond them the white shag carpet of the ice cap stretches to the horizon. No settlements are visible, no signs that anyone has ever set foot there. From twenty-seven thousand feet, it looks like the proverbial last place on earth. Bil Thuma asks, "Can you imagine being an airman who goes down out there and says to himself, 'It's OK, we're going to get out of here'?"

Lou asks me to walk with him to the rear of the cargo bay. Over the thrumming noise of the engines, he tells me that he's annoyed that Jim Blow has only now told him about the new Coast Guard–funded airborne radar survey by the U.S. Army Cold Regions Research and Engineering Laboratory, or CRREL. Lou's also frustrated that CRREL didn't use a device called an airborne magnetometer that he says might have detected metal from the Duck's engine under the glacier.

"We're supposed to be working together," Lou says. "If he had told me, I would have said what we need isn't more radar, it's a magnetometer."

"OK, but it's done. Let it go," I say. "Lou, the new radar gives us more places to search, and the results gave Jim the confidence to convince his superiors to go ahead with this mission. You're on the same side."

"A little communication would be nice," Lou says. "Teamwork, right?"

I stop by Jim's seat, and he hands me the new radar report, complete with maps showing the directional flow of the glacier and the locations described as "strongly prospective targets." Later, when I tell him that Lou would have preferred an airborne magnetometer, Jim reminds me that Lou had been working for months trying to get a private company to run just such a survey. Jim says he expected Lou to be successful on that score, so he focused on getting the Coast Guard and CRREL to provide new radar results, in the hope that the complementary studies would pinpoint search sites two different ways. When Lou's discussions with the airborne magnetometer company fell through, Jim says, he went ahead with the radar because he wanted fresh data.

He has a point: before the results of the CRREL study, most of the radar results were four years old, from what was known as the Essex overflight of 2008. The glacier, and with it the Duck, might have moved since then.

"I think Lou felt left in the dark," I say.

"That wasn't the goal," Jim says. "These results increase our chance of success."

I walk away thinking that Lou and Jim are equally invested in this mission, and also that they aren't seeing eye to eye. More important, I sense that the dustup over the new radar is a harbinger of clashes to come. The pressure they both face to succeed, their different natures, and their conflicting organizational cultures make disputes almost inevitable. If they don't trust one another, or at least communicate effectively, the mission will suffer and the Duck Hunt won't fly.

By default, I have a new role. As the one mission member not formally attached to North South Polar or the Coast Guard team, I hold passports to both. And since I'm on good terms with both Jim and Lou, I can be an honest broker and cultural interpreter. That is, as long as neither shoots the messenger.

WE CAN'T FLY directly to Kulusuk, Greenland, because there's no way to refuel the C-130 at the small airport there. So after midnight local time, we touch down in Keflavík, Iceland, where we'll spend the rest of this night and one more. Even in Iceland, the expedition shopping spree continues. After worrying for days about the crevasse fields, Lou decides that we need an aluminum extension ladder to use as a portable bridge to span deep cracks in the glacier. He sends Alberto Behar to a local hardware store to buy one, then sends Frank Marley to buy another. We'll have to strap them to the skids of a helicopter to get them to Koge Bay. But other than buying propane, gasoline, and diesel fuel in Kulusuk, we're finally ready.

Before we leave Iceland, Lou and Jim gather everyone in a ho-tel meeting room to recount the history of the November 1942 crashes and the mission operations plan. Lou explains that first we'll use global positioning system (GPS) receivers to place marker flags at the six most promising search coordinates, known as points of interest, or POIs. Three of the POIs were spotted during 2008 and are known as Essex One, Essex Two, and Essex Three, the last of which appears to be close to where Balchen placed the X on his treasure map. It's also the most heavily crevassed area we'll be searching. The three other top-priority points of interest come from the new CRREL report, and are known as points A, B, and K. After the six points are marked with GPS, Jaana and the safety team will use the ground-penetrating radar to see if anything that looks like a Duck lies beneath the ice. If not, we'll move on to two more CRREL locations, known as points N and O. If those come up empty, too, it's not certain what we'll do. If there's time, maybe we'll investigate a couple of historic coordinates. If none of those locations reveals the Duck, we'll have to accept defeat and go home.

The next day we fly a little more than an hour to the Kulusuk Airport, where the C-130 kicks up plumes of dust as it touches down on the gravel runway. Mountains and dormant volcanoes dominate the treeless landscape that surrounds the tiny airstrip. Even in August patches of snow crown the nearby slopes.

Seventy miles south of the Arctic Circle, Kulusuk passes for a sizable settlement on the east coast of Greenland, with a general store, a school, about three hundred residents, and twice as many sled dogs. From the air, the fishing village is picturesque, a cluster of homes painted red, blue, or gray, clinging to rocks overlooking a small harbor where stray icebergs take up temporary residence like visiting tall ships. But the impression changes after a mile-long walk from the airport into town, along a winding dirt road that passes two small cemeteries filled with plain white crosses. Up close, most houses are weather-beaten and need repair. The constant threat of frozen pipes means that most have no running water, so the people of Kulusuk, most of them Inuit, haul water from a pumping station and rely on outhouses or chemical toilets. With residents isolated by ice and snow for most of the year, and with little local economy to speak of, dependency on alcohol and welfare is common.

It's a big day in Kulusuk when we arrive, as rifle-toting hunters in small motorboats have brought back two killer whales. The black-and-white whale carcasses are hoisted onto the town's small concrete dock, where an audience gathers as they're butchered, the meat distributed freely among the townspeople. When the townspeople leave, plastic bags with phonebook-sized slices of blubber are left behind, and the whales' severed heads are visible for days in shallow water alongside the bloodstained dock.

The C-130's Coast Guard crew is eager to take off, so they quickly disgorge our gear. Lou climbs onto a giant yellow front-loader as though he's General Patton atop a tank. He hangs onto its side-view mirror to direct its driver as he arranges four tons of equipment, including twenty-five huge Pelican cases, more than a dozen smaller Pelican cases, two Hotsy pressure washers, several ice drills, a five-foot-cube water tank, generators, mounds of duffel bags, and two aluminum extension ladders destined for helicopter airlift to Koge Bay.

THE FOUR-PASSENGER AIR Greenland helicopter lifts off from Kulusuk Airport crammed with five of us aboard: me; Jim Blow; Coast Guard public affairs specialist Jetta Disco; logistics manager Steve Katz; and logbook keeper and aide-de-camp Ryan Sapienza. The Swedish pilot follows the coastline south, occasionally veering from his flight path to show us a two-hundred-foot waterfall at the edge of a cliff, or icebergs shaped like a medieval castle or a slice of cake bigger than a city block. Flying at five hundred feet, then diving toward the water, he circles twice over a bay to point out what looks like a whirlpool of boiling water. Closer inspection reveals that it's a killer whale chasing fish in a feeding frenzy. An hour into the flight, we turn east toward the coastal mountains, flying straight toward the rocky slopes only to pull up and over at what feels like the last possible second. Soon we're cruising over crevasse-covered glaciers toward the North South Polar–U.S. Coast Guard's Duck Hunt 2012 base camp.

We land on the ice forty yards from the campsite, much of it already set up by team members who arrived on earlier helicopter flights. The chosen spot is a cul-de-sac tucked against a rock outcropping called a *nunatak* that overlooks Koge Bay. The rock serves as a windbreak, and the site is no more than one and a half miles from the farthest point we plan to search. Laid out before us on a barren, gently sloped, ice-covered field are a half-dozen orange-and-gray sleeping tents, their stake loops held down by rocks gathered from the *nunatak*. During the next hour, in temperatures hovering around the freezing mark, we'll erect four more sleeping tents to complete our temporary village. Duffel bags, spools of rope, gasoline cans, and half-empty Pelican cases sit in clusters on the edge of the ice field.

Fifty feet from the sleeping tents, at the base of the ridge, is one of two green mini-barn tents, this one filled with mountaineering gear and two Pelican cases. The tent will serve as the command center, and the cases will double as desks to spread out maps to plan each day's assault on the glacier. Atop a rock ledge twenty feet past the mini-barn is the big dome tent, a zippered flap open to reveal several team members drinking coffee, their faces glowing orange from the sunlight through the nylon. On the same ledge, a few feet away, is the second mini-barn tent, filled with food, cooking equipment, and the tireless Michelle Brinsko, getting an early start on lunch. Later, when the beans in her chili refuse to soften, she'll declare that they're a special Icelandic legume called Bierdorff Beans that are supposed to be crunchy. No one who wants to eat another meal dares question Michelle's dubious claim. Her failure-is-not-an-option mandate remains unblemished.

On the other side of the big dome is a fifty-foot-diameter meltwater pond that becomes a reflecting pool at dusk for the setting sun. At first we pump pond water for drinking and cooking, but soon WeeGee Smith and "Doc" Harman drill a hole behind camp that fills with water flowing from inside the glacier. They rig up a pressurized, gravity-fed system that supplies us with water so fresh we could bottle and sell it as Koge Bay Glacier Melt.

On the other side of the rocks is the gray outhouse tent, complete with a small orange flag whose position inside or outside indicates whether it's occupied. "Planting the flag" becomes the camp euphemism for a toilet trip. Lou chooses the highest point of the *nunatak* overlooking the camp for the real flagpole, where he flies the American flag and the "Honor and Remember" mission banner with the names of the three men we hope to find.

Visible in the distance to the west, beyond the *nunatak* and across the Koge Bay fjord, is the sharply rising glacier where the PN9E crashed. Now more than ever, I understand what pilot Armand Monteverde encountered when he steered the big bomber from the end of the fjord over the glacier, only to have the ice rise up to meet him. Seventy years later, the glacier where the B-17 crew awaited rescue remains laced with crevasses. The fjord is jammed with icebergs calved from the glacier's two-hundred-foot-tall leading edge. The thunder of new icebergs being born echoes through base camp.

A VIEW OF THE DUCK HUNT CAMPSITE FROM THE *NUNATAK* OVERLOOKING THE TENTS. *(MITCHELL ZUCKOFF PHOTOGRAPH.)*

The weather forecast is clear and mild for the workweek ahead, with temperatures in the teens at night and the high forties during the day. Best of all, there's no sign of polar bears, rendering the shotgun and rifle wise but unnecessary precautions. The Snublebluss warning system that Lou ordered from England never arrived, so the absence of bears is especially good news. I compliment Frank Marley on how smoothly the camp setup has gone, but he's not impressed. "Camping is the easy part," Frank says. "That's not what we're here for."

DURING THE PAST few months, Jim Blow has grown intrigued by the X on Bernt Balchen's 1943 hand-drawn map. Our base camp is less than a mile from Blow's best estimate of the site of Balchen's "X." The exact location can't be known because Balchen didn't include coordinates, and his ink and watercolor drawings captured some, but not all, of the coastline and the rock ridges that rise from the glacier. The sketches also simplify the Koge Bay area by omitting perhaps sixteen miles of glacier-covered land between the X and Comanche Bay, a geographic flaw that long raised questions about the map's accuracy.

To gain a better perspective, Jim, Lieutenant Commander Rob Tucker, and I climb to the top of the rocks at the edge of camp. We imagine the flight path John Pritchard might have taken in the Duck from the PN9E over Koge Bay to the spot on Balchen's map. It's at once easy and chilling to envision the little biplane rising from the bomber's crash site and sweeping over the glacier, then turning toward Comanche Bay, into whiteout conditions that would render an ice-covered hill an invisible and deadly obstacle. Jim's mind remains fixed on Balchen's map.

"Absent a photograph of the actual crash site," he says, "that is the best piece of data that we have because it was drawn by an eyewitness. The more study we've done of the charts and the satellite images of this area, the more that Balchen's 'X' basically is our greatest point for re-creating what happened that fateful day. Based on that and what we know about the ice movement, and the lay of the terrain now, Balchen's 'X' has a lot more credibility as we're out here."

Jim has an explanation for the maps' missing swath of land. "Now I'm thinking he omitted that on purpose, because that's a lot of detail that didn't need to be in there. He was just trying to show where the aircraft was."

Later, Jim and I hike a half mile to the end of the *nunatak* to look over the crevasse field that contains the Essex Three/Balchen X sites. Returning to his pilot roots, Jim balances on the rocks and imagines how John Pritchard might have ended up there. "He's coming down the fjord, he's calling for magnetic orientation, trying to get a bearing to the ship." Jim points to a rock outcropping at the edge of the water. "I would estimate that he would see this upper cliff, and as soon as he thought he was clear, he's going to turn to the bearing that the ship gives him, which I assume that he has heard."

Jim pulls out his iPhone and uses a digital compass to find the direction to Comanche Bay. The compass bearing from where we stand is 115 degrees true, or exactly the course the *Northland* tried to give Pritchard. "At 115 degrees," Jim continues, "he would turn and be going in that direction, which is into that ridge"—a snow-covered hill several hundred feet high—"which he would not be able to discern. I'm just curious, if he hits here, bearing 115 into this ridge, where would he be on this slope seventy years later?" Jim is equally pleased that the area fits witness descriptions that the crash site was an open field sloping half a mile to the water.

The possible value of Balchen's X is one subject on which Jim and Lou agree. Lou has been telling me for months that the Essex Three point of interest aligns with Balchen's X and holds special allure for him. Yet both men know that the under-ice anomaly spotted by airborne radar at Essex Three might be nothing more than a crevasse or pooled water atop bedrock. Also, even if Balchen was correct in 1943, the glacier's movement might have carried the Duck's wreckage a half mile or more, to the vicinity of Essex One, Essex Two, or the new CRREL target points A, B, and K.

The bottom line is that unless Jaana's ground-penetrating, boots-on-the-ice radar reveals a powerful image of an anomaly roughly the size of a crushed Duck, all the planning and speculation in the world won't matter.

ALMOST FROM THE moment she stepped off the helicopter, Jaana has been linked by a rope umbilical cord to safety team members, most frequently John Bradley and Frank Marley. Their first task has been to use a handheld GPS receiver that John brought with him to translate the mission plan's points of interest from latitude and longitude coordinates to actual places on the glacier. When the coordinates on John's GPS match those of a POI, the team plants an orange flag at the spot, as a reference point for when they return with Jaana's radar to sweep the area.

When safety team member Nick Bratton joins the flag-placement teams, the size and conditions of the Koge Bay glacier worry him. He writes in his journal, "My first foray onto the ice was a cautious one. I've spent lots of time on lots of big glaciers, but this was on a different scale altogether. Words cannot describe how incredibly enormous this glacier was. From our camp the ice stretched to the north and west as far as the eye could see." Nick is troubled that our campsite is in a transition zone, an unpredictable place where snow is in the midst of becoming glacial ice. He recounts his first survey trek to his journal: "Most crevasses were visible but filled with snow, so the depth and strength of the snow bridges were unknown. As for the unseen crevasses, who knows how many there were? Those are always the real danger. You rarely fall into a crevasse you can see. The hidden ones are what get you."

After flagging the first few points of interest, Jaana begins her radar work in earnest during her second day on the glacier, Friday, August 24. First on her list are Essex One, Essex Two, and Point K, which is between those two Essex points. Dragging the dragon-tail antenna behind her, she clomps back and forth over each site, watching the screen on a control unit that hangs from a harness at her waist. If she notices any anomalies, she can circle back for another look. The control unit also records the radar readings and location coordinates for analysis back at base camp.

Jaana explains that the radar takes a cone-shaped view of the glacier down to the bedrock, more than a hundred feet below. Any interruption in the solid ice appears on the screen as shapes called hyperbolas, which to the untrained eye resemble flatter versions of McDonald's golden arches. One problem is that crevasses create hyperbolas similar to ones made by, say, the wreckage of an amphibious World War II Coast Guard biplane. Jaana's expertise enables her to distinguish between a crack and a Duck.

As she scans the first points of interest, Jaana adjusts her route to avoid surface crevasses, but she and her safety minders try to walk in lines as straight as possible over the crunchy snow, to create overlapping cones of radar and to create a complete image of what lies beneath each point of interest.

Anticipation rises when Jaana, Frank, and John return to base camp for lunch. Inside the dome, Jaana transfers the data to her laptop computer, but nothing on her screen from the first three points of interest looks like an airplane. These were among the most promising sites, and the bad news spreads. Disappointed expedition members wander in and out of the dome, hoping that Jaana finds something unexpected during her closer look. When she doesn't, Lou is deflated. "I feel like walking out there and seeing where the Duck might have flown into the glacier," he says. "I want to appeal to the spirits of Pritchard, Bottoms, and Howarth, to ask that they help us find them."

Nick's concerns about safety grow when he sees Jaana's data. Under what appears to be solid stretches of glacier ice, the radar shows fractures deep and not-so-deep beneath our feet. "Shit like that worries me," Nick tells his journal. "There is no way to mitigate that. Walk fast."

As lunch ends, whispers race through camp that there's a problem with our GPS receivers, which means that the flag placements for the points of interest are in the wrong places. On one hand, the good news is that if Jaana repeats the radar survey in the right locations, she might get better results. But a bigger worry emerges: if the GPS receivers don't work, we can't know where the flags properly belong. A mistaken GPS reading that places a point of interest flag as little as thirty yards

JOHN BRADLEY LEADS THE RADAR TEAM ONTO THE GLACIER, FOLLOWED BY JAANA GUSTAFSSON, WITH THE GROUND-PENETRATING RADAR DEVICE, AND FRANK MARLEY. (MITCHELL ZUCKOFF PHOTOGRAPH.)

from its correct location might be enough to ruin the search. In that case, Jaana might as well take a drunkard's walk across the glacier, randomly taking radar readings. That approach would have about the same chance of success as shooting an arrow into the air blindfolded and expecting it to pierce an apple on Lou's head.

Tensions rise as word of the GPS problem spreads. Jim fumes that we've been relying on small personal GPS devices, yet two highly accurate Trimble-brand GPS receivers sit idle in the dome tent. "We're talking about mission-critical technology, and no one knows how to operate the Trimbles," Jim says.

Lou says he thought that Jaana or Bil would know how to use the Trimbles, but it turns out neither does. At the moment, we're forced to rely on unreliable personal GPS units. For instance, John Bradley's GPS unit places Essex Two on a rocky slope instead of where everyone agrees it belongs, in a wide-open area of the glacier.

The GPS issue compounds Jim's irritation that no one from North South Polar knows how to operate a metal-detecting magnetometer that Lou brought. The magnetometer job falls to Terri Lisman, the geophysicist from the National Geospatial-Intelligence Agency, who's part of the Coast Guard team. Terri gamely reads the manual and assembles the unit. She follows Jaana to the erroneous points of interest and, not surprisingly, detects no metal under the ice.

I find Lou near the kitchen tent to be sure he's aware of Jim's exasperation. He says he's already on top of the issue. "If we don't have GPS," Lou says, "we've got nothin'." He asks Alberto Behar to apply his PhD in electrical engineering to the task of figuring out the Trimble units. Lou considers this par for the course on such a complex mission. "It's easy to point fingers and blame. The crew of the PN9E could have done the same thing. But they had to survive, and we have to succeed. They worked together. I wish

Jim had come up to me and said, 'Lou, how do we work together to fix this?' "

Although Lou is the mission's civilian leader, frustration among the North South Polar and Coast Guard team members congeals around Steve Katz, Lou's second-in-command, whose job is to direct personnel and keep the mission on track. Privately, expedition workers complain that Steve seems unfamiliar with the mission plan, issues confusing and sometimes contradictory orders, and appears unsure how to address the GPS crisis. For his part, Steve says his army experience has taught him to expect grousing from the ranks when things get rough.

Yet soon I can't walk more than a few feet without someone pulling me aside to criticize Steve's leadership and preparedness. I wonder if team members' widespread doubts are by themselves evidence that Steve is outside his element and is losing control of the troops. Still, he seems untroubled, and later he relaxes alone on the *nunatak*, smoking a cigar.

As frustration deepens over the inaccurate GPS readings, nerves get frayed, patience wears thin, and an old word gets a new meaning. In its original use, an icehole is a shaft that WeeGee melts into the glacier using the Hotsy. Or, more accurately, it's a shaft that he *would* melt if we knew where to search. As hours slip past with little progress, *icehole* is repurposed as a term of disparagement, used to describe anyone seen as not carrying his or her weight or doing his or her job. As in, "He's being an icehole." The phrase gets a workout.

After lunch, Jim and Steve hike up the glacier toward Essex One and Essex Two to visually compare the flags' GPS-dictated locations with the points as they appear on the mission map. While they're out, Bil Thuma calculates the distance between base camp and the flags' proper locations, to determine how far off the GPS-determined placements might be. By nightfall, when Bil's calculations are complete and everyone is back in camp, the expedition's leaders reach a dismaying conclusion: the flag placements are up to several hundred yards from where they belong. Worse, with two days gone and less than five days remaining, no one knows how to fix the problem.

I pass WeeGee on the way to my sleeping tent.

"So, Mitch," he asks, "how does it end?"

21

CROSSED WIRES

THE WIND WAS screeching, the snow was blowing, and temperatures on the glacier were subhuman. Armand Monteverde and Paul Spina knew that if their delusional companion Clint Best remained outside their snow cave, death would come quickly. Yet they feared that if they followed him into the pitch-black night, they might die alongside him.

Spina urged Monteverde to shine the flashlight around the tunnel entrance, so Best might see the beacon and follow it back inside. Monteverde moved to the doorway with the light, but before he reached the entrance they heard a crash. Unable to see in the dark, Best had somehow tripped and fallen headfirst back into the tunnel. His companions rushed to him and saw that his face and hands were a deathly shade of blue. Monteverde pulled Best inside and, with Spina's help, wrapped him in a sleeping bag.

Half afraid and half angry, their emotions overflowing, Monteverde and Spina yelled at Best, and he began to cry. Still gripped by delusions, he told them that he'd gone outside to get their car, because neither of them had had brains enough to move it into the garage before the engine froze. It was hard for Monteverde and Spina to stay mad at him after that.

The pilot and the engineer stayed up all night with Best, praying that Pappy Turner would fly over in the morning with supplies, advice, and news about the next rescue attempt. Turner's B-17 did come, but it made only one supply drop and flew off before they could connect on the walkie-talkie. Monteverde uncovered the entrance to the tunnel and found a package just feet away, as though Turner and his crew were milkmen who delivered to the front step. Inside were rare treats: roast beef sandwiches, cookies, candy, and toothbrushes with paste. Monteverde and Spina thawed out the sandwiches and brushed their teeth for what seemed like hours. Best's mental state remained unchanged. Still unhinged, he wouldn't eat. They took turns watching him all day and night.

Turner flew over again the following day, and this time Monteverde and Spina described Best's breakdown over the walkie-talkie. Turner said he'd speak with a medical officer at Bluie East Two and return with medicine and instructions. Before leaving, Turner dropped bacon, ham, candy, and cigarettes. One package also included candles, which they needed to watch Best at night.

The medicine arrived as promised. Antipsychotic drugs were a hit-and-miss proposition in 1943, so the pills Turner dropped might have been barbiturates, widely used at the time as powerful sedatives and anticonvulsants. After taking the pills, Best slept for about four hours and stopped shaking and sweating. When he woke, he looked and seemed more like himself. Best wolfed down bacon sandwiches and seemed to reclaim his right mind.

That night, Best offered to reprise his role as camp cook. All was well at first as he prepared their meal, but then he grew quiet and unnaturally still. Monteverde called to him but got no reply, so he tucked Best into his sleeping bag. Aware that he was sinking into a new delusion, Best told them that "his wires were getting crossed again," as Spina put it. He asked for another pill, which put him to sleep.

WITH BEST UNRAVELED and Spina's arm still ailing, Monteverde was the only one who could look for the supply plane or rid their hole of trash and waste. The heavy snows of February piled on top of their quarters, turning the tunnel passageway to the surface into a narrow, icy chimney. One day while trying to shimmy up to go outside, Monteverde became stuck halfway, his arms pinned to his sides and his ice-logged clothes making it impossible to wriggle free. Best couldn't rouse himself to help, so Spina used his good arm to grab one of Monteverde's thrashing legs and pull him back down.

Another time, Monteverde was gone awhile, so Spina poked his head outside to look for him. Fear swept over Spina as he saw nothing but an empty, endless glacier. He yelled for Monteverde but heard no answer. Verging on panic, Spina thought that Monteverde had fallen into a crevasse. Roused from sleep by Spina's yells, a now-clearheaded Best told Spina not to go outside. Best knew that Spina had spent most of the past three months inside the PN9E tail or the snow cave, so he might not recognize hidden crevasses. As Best got dressed to search for their leader, Monteverde appeared in the tunnel entrance. He explained that he'd gone down into a sunken ice bridge to retrieve a can of precious kerosene for a new stove that Turner had dropped. He said he didn't answer Spina's cries because he feared that Spina would hear his voice and think it was coming from a crevasse, then race outside to rescue him.

After that, Monteverde and Spina developed a routine in which Monteverde went outside to haul in the packages to the entrance and Spina pulled them into the cave with his good hand. As weeks passed, the food supplies became plentiful and more elaborate, with deliveries that included a dozen roast chickens, pork chops, and cooked steaks. One night when Monteverde and Spina dove in to a chicken feast, Best told them he didn't feel well and needed another pill. Again it put him to sleep.

When the weather eased in early March, Turner's B-17 returned bearing a natural remedy for Best's troubled mind: letters from home. When they learned by walkie-talkie about the incoming mail, all three men went outside to watch as Turner circled overhead, lining up for a low and careful drop of the irreplaceable cargo. They pounced on the package as soon as it hit. Best got the lion's share, but there were letters for all three. Each man read them again and again, sharing the best parts with the others. A favorite passage came in a letter from Monteverde's family. It said the War Department wouldn't reveal where his plane crashed, but the family pieced together enough clues to conclude that he was stuck in a place with lots of ice. Monteverde's relatives in sunny California advised him to be patient and wait for it to melt. The trio laughed at the thought, knowing that if they followed that advice they'd be stuck in Greenland for several millennia.

Whether a result of the pills, the letters, or something else, Best climbed from the depths of despair and regained his hold on reality. He continued to feel downhearted, but he no longer showed signs of being delusional.

Life settled into a tedious routine in the PN9E cave. The men had enough food and an endless supply of snow to melt for water. They had reading material and candles to read by. They read aloud, each man taking a turn until his hands ached from the cold while holding a book or magazine outside his sleeping bag. Then he'd hand off to the next man in their reading circle.

Their favorite selection was an essay in *Reader's Digest* about the power of prayer. One section seemed especially apt for their predicament:

It is the only power in the world that seems to overcome the so-called "laws of nature"; the occasions on which prayer has dramatically done this have been termed "miracles." But constant, quieter miracles take place hourly in the hearts of men and women who have discovered that prayer supplies them with a steady flow of sustaining power in their daily lives.

The essay renewed their hope, and they read it again and again.

ON A DAY clear enough for a supply flight, Pappy Turner told the survivors that Bernt Balchen and Barney Dunlop would try another rescue attempt with a PBY Catalina. The plan called for the rescuers to land briefly near the now-empty Motorsled Camp. Then, a dogsled team would snake its way through the crevasse field to the PN9E crash site and guide or carry the three remaining crewmen back to the Motorsled Camp. The PBY would then fly them all to Bluie East Two.

Several weeks of storms intervened, and the PN9E trio went for a long stretch without fresh supplies. Temperatures fell and winds raged at up to 125 miles per hour. The three men had a cache of food, but they wouldn't leave their sleeping bags to cook or even thaw it. They lived on concentrated chocolate bars. Their candles ran out and their sleeping bags froze.

When Turner flew overhead after the storms eased, he told the survivors that some of the men at the base had bet that Greenland's weather had finally killed them. It sounded heartless, but he meant it as praise for their endurance. Turner rewarded their stubborn refusal to die with gallons of strawberry jam that Spina had requested; new sleeping bags; three air mattresses; and a bottle of whiskey. This time Spina took sips, not gulps.

On March 17, 1943, the planners at Bluie East Two concluded that the weather was clear enough to try a second Dumbo-on-ice landing. Along for the trip with Balchen, Dunlop, and their volunteer crew were the trail team of Captain Harold Strong and sergeants Joseph Healey and Hendrik "Dutch" Dolleman, three men who'd come to Greenland with a wealth of experience on ice.

Healey had grown up in Boston's Irish-American enclave of Dorchester. Tall and strong, as a teenager he'd been a second mate aboard the supply ship *Jacob Ruppert* during Admiral Byrd's 1933–1935 Antarctic expedition. Healey joined Byrd again for a 1939–1941 polar journey, this time as a dog wrangler. During the latter expedition he received a singular if obscure honor: on the east coast of Antarctica, in a place called Palmer Land, on the north side of the entrance to a body of water called Lamplugh Inlet, lies a square-shaped outcropping of rock named Cape Healey. Six months before the attack on Pearl Harbor, Healey left Byrd's service and joined the U.S. Army as a sled dog trainer.

Dolleman was born in the Netherlands and immigrated to the United States with his parents. After a boyhood in Manchester, New Hampshire, he joined the U.S. Army, which sent him to Antarctica on a scientific expedition to study penguins. Short and lean, days away from his thirty-seventh birthday, Dolleman also was the namesake of a little-known land mass. Off the east coast of Palmer Land in Antarctica lies the ice-covered, thirteen-mile-long Dolleman Island. After his army-sponsored penguin studies, Dolleman was sent to Greenland.

Of the three, the most exotic backstory belonged to Strong. A native of Gloucester, Massachusetts, the forty-year-old Strong graduated from Princeton in 1924, then spent two years studying art and architecture in Europe. He returned to the United States in time to make a killing in the stock market, cashing out before the 1929 crash. Flush with cash, Strong and a friend went to Alaska to herd reindeer for two years. On a trip home they saw the ongoing effects of the Great Depression, so they returned to Alaska and worked for five more years in the 1930s as fur traders, buying fox pelts from Inuit hunters along a twelve-hundred-mile circuit they traveled by dogsled. Later, incongruously, Strong worked as a wallpaper executive in Texas before growing bored of civilized life. He joined the Army Air Forces in 1942, and his peculiar expertise earned him a spot on an Arctic search-and-rescue team in Greenland. Muscular, tanned, tall, and square-jawed, Strong cut a dashing figure as he cruised around Bluie West One in a Jeep pulled by a team of huskies. For a valet, Strong employed a huge black Newfoundland dog equipped with saddlebags to carry his lunch and his holstered .38.

FLYING SOUTH FROM Bluie East Two, Barney Dunlop lined up to land the PBY Catalina one hundred yards to the right of the Motorsled Camp. As he descended, the clear weather several thousand feet above Greenland became what Balchen called "an opaque sheet of driving snow particles, whipped up by the wind. Our visibility is no more than fifty feet, and the blowing snow is so thick that we can barely see our tip floats."

Dunlop fought the elements and slid the Dumbo to a stop about five hundred yards from where he'd intended, but in one

FROM LEFT, SERGEANT JOSEPH HEALEY, COLONEL BERNT BALCHEN, CAPTAIN HAROLD STRONG, AND SERGEANT HENDRIK "DUTCH" DOLLEMAN. *(U.S. ARMY PHOTOGRAPH.)*

piece. The three-man trail party climbed out with their dogs, sleds, and gear. Before heading toward the PN9E, they rocked the plane free from the glacier's grasp. With the windshield coated by ice and winds toying with the rudder, Dunlop returned to the air with little more than a prayer to guide him. Soon he cleared the weather and returned to the base.

Strong, Dolleman, and Healey could only manage part of the six-mile journey before darkness fell, so they camped that night on the ice. The following day, Turner's B-17 flew one pass after another to guide the dogsled team toward the PN9E wreck.

Monteverde, Spina, and Best knew from walkie-talkie chatter that the rescue team was approaching. They climbed up their igloo's ice chimney and made a smoky oil fire on the glacier to serve as a beacon. They made extra coffee and assembled ham sandwiches, as though waiting for neighbors to drop in for lunch. The three survivors stared into the blinding white distance as tiny black specks grew larger. They grew impatient as long stretches passed without any apparent progress. Going was slow, as the men on the trail team avoided crevasses and repeatedly lifted their sleds past waves of *sastrugi*. Several times the sleds overturned as they hauled them through deep drifts, tiring and slowing them further.

Spina grew impatient and climbed atop the front end of the PN9E like a hood ornament. As he waited for the trail team, he spotted a can of milk some fifty feet away. Without thinking, he climbed down and walked across the glacier toward it. When Spina reached the can, he noticed a small crack in the ice. He stared down at it, mesmerized by the jewel-like blue-green translucence. Monteverde inched in his direction and told Spina that he was standing on a bridged-over crevasse. If it collapsed, he'd be enjoying the lovely view for the rest of his short life.

Chastened, Spina backed away. Safe inside the snow cave, he got an earful from Best for risking his life—without retrieving the precious milk.

Hours passed, and finally the survivors could hear sled dogs barking and the trail team calling out commands. For a time, rescuers disappeared from sight in the dip of a small ravine. Monteverde, Spina, and Best called to them for a half hour but received no answer. Fearing the worst—three more men lost in a crevasse—they slouched back into their cave and broke down in tears. The dread they'd felt that more men would die on their behalf, a motive for their brief suicide pact, had apparently come true.

As they wept, the trio heard someone walking on the bomber wing that served as their roof. Spina heard a voice call, "I guess these ice worms don't want to leave this joint!" The survivors rushed outside and shook hands with Strong, Healey, and Dolleman. The trail team explained that they'd disappeared in a gully a half mile away to pitch camp and to chain their dogs out of range of the crevasses. The men got to know one another, drank coffee, and ate ham sandwiches. Then the trail team returned to their dogs.

Back in the ice cave, the survivors talked about how little exhilaration they felt at the rescuers' arrival. Spina suspected that their emotions, like their bodies, had been dulled by the cold. He wrote in a journal, "After 129 days of fighting everything the Arctic could throw at us, I guess nothing could excite us."

That night, the PN9E trio unwrapped clean, dry socks and fresh clothes they'd saved for their departure. They'd been unwashed and unshaven for more than four months, their hair so long and unkempt that they barely resembled their fresh-faced military portraits. They reeked, and they knew it. But at least when they left their Greenland snow cave, they'd be wearing clean underwear.

They climbed into their sleeping bags but couldn't sleep, so they talked all night. Spina spent the next morning vomiting, either from too much ham or too much bottled-up anticipation. While Spina emptied his stomach, Best dug up the PN9E's Norden Bombsight, which they'd buried in ice for safety. Spina pulled himself together, and the three men carried out their sworn duty, destroying the bombsight by throwing it into a crevasse. They listened for it to hit bottom, but they never heard a sound.

When the trail team arrived, it was time for Monteverde, Best, and Spina to abandon ship. Before leaving, Spina suggested that they burn the PN9E's cockpit, to keep its secrets from enemy eyes. Strong told him not to worry; no one would ever be brave, foolish, or unlucky enough to come this way again.

22

THE TEN-METER ANOMALY

AUGUST 2012

IN THE PREDAWN hours of Saturday, August 25, Jim Blow hears what sounds like frozen rain drumming against his tent. The noise takes him back two years, when the last attempt to find the Duck was nearly washed away by relentless downpours. Burrowed inside his sleeping bag, Jim turns to Rob Tucker, his tent mate and fellow Coast Guard pilot: "Rob, this is where the shit hits the fan." But at daybreak they realize that the sound was granular snow beating against the tent's nylon sides. Clear skies are holding on the Koge Bay glacier, at least for now. But good weather is meaningless if we don't know where to search.

After Spam on bagels for breakfast, science and safety teams head out from base camp hoping that the GPS issue has been resolved and the receivers are working properly. But by noon, Jim is smoldering: "We still can't verify the points of interest." He leaves camp, hiking more than a mile toward the rocks closest to the water. Ostensibly, he hopes to find scrape marks on a ridge if the Duck hit a *nunatak*, or pieces of the plane deposited by the flow of the glacier. He knows that the chances of success are slim to none, but it's better than sitting idly in the dome tent, nursing coffee and aggravation.

Jaana continues her radar work, roped together with Frank and John to again sweep the best-guess locations of Essex One, Essex Two, and Points A, B, and K. Terri straps the magnetometer onto her back, its mast antenna extending like a flagpole four feet above her head, and goes with Nick and Alberto. Nick writes later, "Terri stomps back and forth across Point O, mumbling to herself about the readings and the data not saving. Basically, the instrument wasn't calibrated, she didn't really know how to use it, and those issues could have been resolved in camp instead of wasting our time on the glacier. Oh well, another pleasant walk out on the ice."

Although the magnetometer's effectiveness is suspect and the GPS coordinates remain dubious, the work continues under the theory that maybe the flags are close enough to the points of interest for the equipment to spot a large metal object embedded in the ice. Also, there's nothing else to do. With our days on the glacier numbered, the only alternative would be to give up.

Both teams return to base camp hours later, tired and empty-handed. "I really was not seeing anything," Terri says. "It was flatline the whole way."

Throughout the morning, Lou's been furiously making calls on his satellite phone, seeking answers everywhere from the Kulusuk Airport control tower to Colorado, where he reaches a geophysicist from the 2010 expedition. The call evolves into a lesson that helps Alberto to program the high-tech Trimble GPS receivers. From a separate call, Lou learns that an Air Greenland helicopter is en route with a technician to service a nearby ground station that enhances the accuracy of GPS satellites. GPS ground stations serve as fixed reference points; when they work correctly, they allow GPS receivers to pinpoint latitude and longitude coordinates within inches. It's not clear what service the ground station needs—a reboot is one guess, a swift kick is another. But Lou thinks that the ground station might be the culprit for our problems, and when it's fixed the GPS units will work properly.

Jim isn't buying it. He dismisses the service visit to the GPS ground station as an unrelated coincidence. Instead he blames human error, caused by Lou and Steve's mistaken assumption that Jaana or Bil would know how to operate the Trimbles. Because they're unfamiliar with the devices, we're forced to rely on the personal GPS units brought by John and two Coast Guardsmen, Rob Tucker and "Doc" Harman, which are putting out misleading readings.

By Saturday afternoon, after days of trial and lots of errors, Alberto and several others realize that the personal GPS units were programmed to the wrong setting. That setting, Jim says, "works great for the United States, but can be off by as much as four hundred to five hundred meters in southern Greenland." He's certain that the mistaken settings are the real reason for the misplaced flags.

The personal GPS units are reset for Greenland, and Alberto loads the Trimbles with the mission's points of interest. Tests show that all the units—the personal devices as well as the Trimbles—now display the same readings for the same locations. Best of all, when the units are checked against the known coordinates of the base camp, it's a perfect match.

"Mistakes happen," Lou says. "It's how you fix them that counts."

On our third day on the glacier, we're in business. Just in time, too, as relations between North South Polar and the Coast Guard have deteriorated. One exchange between the teams' leaders outside the command tent captures the mood.

"We're almost starting to plan the evacuation," Jim tells Lou, "and we haven't done much of anything."

"What do you want me to do, Jim," Lou asks, "just start digging random holes in the ice?"

"No, Lou," Jim says, barely containing his anger. "I want you to dig six fucking holes where I told you to dig six fucking holes, at the points of interest."

After more discussion, the expedition's leaders walk off in opposite directions.

LATE THAT AFTERNOON, with a fully loaded Trimble in hand, Nick, Alberto, and I suit up to return to Essex One, Essex Two, and Points A, B, and K, to place the flags in their correct locations. Before we leave base camp, Nick and Alberto argue over whether we need to be roped together on the glacier.

"I've been on expeditions where the work doesn't get done because of 'safety' getting in the way," Alberto says. Nick says his wife wants him to return home alive, so we'll use ropes. Alberto, who has a wife and three children, tells Nick that he resents the suggestion that he's cavalier about safety or his family's interests. He points out that repeated treks to the area where we're headed have proved that the glacier there is stable, and we'll be able to see the crevasses along the route. Nick holds his ground, arguing that a few minutes to rope up would be insignificant compared to the delays we've already faced, and the small investment of time might prevent a crisis that would doom the mission. Soon we're roped together in a line, with me in the middle, twenty-five yards behind Alberto and twenty-five yards ahead of Nick.

The hardened snow crunches beneath our boots as we head northeast from base camp. The sky is a brilliant aquatic blue. Sunshine bounces off the ice. We hike up and down moguls and drainage channels, occasionally coming across areas as smooth as a skating rink. There's no sign of life anywhere: no plants, no birds, no animals, no insects, just ice. Nick and Alberto don't make small talk, so the glacier is quiet except for the wind, our footsteps, and our breathing. When we reach a seven-foot-wide ice bridge spanning a crevasse, I take two fast steps across, then silently celebrate that Nick prevailed on the safety ropes. If I fall through, I tell myself as I cross, they'll pull me up. The bridge holds firm and we keep going.

We reach the first point of interest after twenty minutes and spend the next hour moving the orange flags, writing the date and the Trimble-supplied coordinates on each. Most of the new placements are about one hundred yards from their previous positions, far enough to have put Jaana's initial radar sweeps out of range of the most promising anomalies. Before returning to base camp, we look across the glacier and see the flags arranged in exactly the pattern the mission map predicted.

JAANA GUSTAFSSON DISPLAYS A COMPUTER IMAGE OF THE RADAR LINES SHE HAS WALKED ON THE GLACIER WITHOUT GETTING A RESPONSE SHE THINKS WOULD COME FROM THE DUCK. *(MITCHELL ZUCKOFF PHOTOGRAPH.)*

The correctly plotted points of interest spark a wave of enthusiasm. The feeling intensifies when Jaana runs her radar over the
newly confirmed flag locations and reports possible anomalies near
Points A and B. Nothing turns up at the other spots, but at least
now we have something to melt toward.

That night, I sit next to Jaana in the dome as she reviews the
data on her laptop. She's quiet by nature, with a dry sense of humor.
It's obvious that she's not excited about the readings. She shows me
her computer screen and explains that although the hyperbolas near
Points A and B depict something under the ice, they seem likely to
have been made by empty voids or crevasses. There's a chance they
might be the Duck, but she wouldn't bet on it. They're just the best
options she's seen so far. I slink to my tent and crawl into my sleeping bag, the raw discomfort of the ice sheet beneath my back fitting
my mood. My only consolation is that WeeGee didn't have a chance
to ask me tonight how the book ends.

ALBERTO BEHAR LEADS THE MAGNETOMETER TEAM FROM CAMP, FOLLOWED BY TERRI
LISMAN, WITH THE DEVICE STRAPPED ON HER BACK, AND W. R. "BIL" THUMA. *(U.S.
COAST GUARD PHOTOGRAPH BY JETTA DISCO.)*

THE NEXT DAY, as everyone eats breakfast in the dome, Steve tries to pump us up: "This is probably the most critical day of our little jaunt out here, and hopefully it will be a success." In a flurry of activity, Jaana, Nick, and John head out to Point O, while Terri, Bil, and Alberto take the magnetometer to survey the anomalies that Jaana identified around Points A and B. The rest of us fall under WeeGee's command as members of Team Hotsy.

Compared to the expedition's other critical pieces of equipment—the radar, the magnetometer, the Trimbles, and Alberto's down-hole video camera—the Hotsy is a dancing bear among ballerinas. The seven-hundred-pound pressure washer is the size and shape of a basement freezer. It sports a red steel frame on twelve-inch tires; a silver, keg-sized burner filled with heating coils; a Honda engine; tanks for gas and diesel; a car battery; a muffler that doesn't prevent it from sounding like a jet engine; and assorted valves and hoses. One hose draws water into the Hotsy, which heats it to 225 degrees Fahrenheit. The boiling water is then spit out through another hose at 3,500 pounds of pressure per square inch. In civilian life, the Hotsy's purpose is high-powered cleaning, with enough force to tear barnacles off boat hulls and graffiti tags off buildings. Here its job is to interrogate the glacier until it surrenders its secrets.

To do its job, the Hotsy must be stationed above the site of a radar anomaly, so WeeGee can melt down to check it out. The problem is that the shortest distance between base camp and Point A is a half mile uphill, across a crevassed glacier. Ideally, we'd call for an Air Greenland helicopter to airlift the machine, but none is available. Our only option is brute force. WeeGee assesses the situation and decides that in a match between the Hotsy and us, we'd lose. So he changes the odds. He detaches the two halves of one of the extension ladders bought in Keflavík, then straps one half across the front of the Hotsy and the other half across the back. With two ladders extending parallel to one another from the body of the machine, the Hotsy bears a squinting resemblance to a certain World War II amphibious biplane.

By attaching the ladders, WeeGee has turned the Hotsy into a huge blocking sled. Nine people—four in front and five in back—can grip the ladder rungs and the Hotsy's frame to push in unison. At the sight of this innovation/contraption, Coast Guard public affairs specialist Jetta Disco coins a slogan for whenever problems arise: "Just ask yourself, 'What would WeeGee do?'"

Team Hotsy gathers along the ladders as WeeGee directs the ascent to Point A. While the rest of us assume our pushing positions bundled in heavy layers, hats, and gloves, WeeGee works gloveless and hatless, in black snow pants, a black sweater, and his trademark orange boots. Koge Bay doesn't qualify as cold to WeeGee, who spent months carving *Glacier Girl* out of the ice. A consensus emerges that after so much time in Greenland, he's now half polar bear.

We get rolling to cries of "Let's melt some ice!" The combined North South Polar–Coast Guard effort gains steam. We're almost trotting as we move up the steady incline. It's not an ideal moment, but during a rest stop I tell the story of Clarence Wedel falling through a hidden crevasse seventy years ago, a few miles away. Fortunately, the glacier allows us to remain on its surface, even as we push the Hotsy across at least five bridged-over crevasses to Point A.

While the rest of us catch our breath, WeeGee fires up the engine, arranges the hoses, drills a hole to create a water source, and gets to work. Normally, the Hotsy uses a spray gun at the end of the hot-water hose. But WeeGee replaces it with a ten-foot black steel pipe. On its end he screws a pointed silver nozzle that pushes against the glacier as it melts a six-inch-diameter hole. It takes forty-five minutes or more to reach a depth of fifty to sixty feet. While WeeGee works, there's little for the rest of us to do but wait and hope that he hits something.

With countless details to manage, Lou returns to base camp. While looking through his bags, he finds a gold ball that's supposed to screw into the top of the flagpole. He cradles it in his hands and waits for help mounting it. Jetta tells Lou that military lore suggests that a hollow flagpole ball should contain a match and a bullet. A commander facing defeat is supposed to burn the flag to prevent its desecration and use the bullet to commit suicide.

AFTER MELTING SEVERAL holes at Point A, and instructing me in the little-known art of Hotsy glacier drilling, WeeGee drops Alberto's camera into the holes, hoping to see signs of Duck. The camera's images are projected onto a twenty-inch diagonal video screen Alberto has embedded in a small Pelican case, and we crowd around to watch. The images resemble a colonoscopy in an icehole, although here everyone hopes that the camera will spot a foreign object. But one hole after another contains nothing but ice.

That wouldn't bother WeeGee—the radar can be imprecise, so maybe whatever caused the anomaly is a few feet from the holes. Under normal circumstances, he'd simply melt more holes. But WeeGee notices a disturbing pattern: water from the Hotsy pipe builds up as though stopped by a clogged drain. But when each hole reaches a certain depth, the water disappears. It's a mystery to the rest of us, but to WeeGee it's bad news: the holes are draining into an underground ice void. The anomaly that Jaana spotted isn't the Duck. As she suspected, it's a subsurface pocket of air in the glacier. WeeGee delivers the disappointing news to base camp via walkie-talkie. The other anomaly proves no different. We leave the Hotsy on the glacier and trudge back.

LATER THE SAME day, on the ice beyond our sleeping tents, WeeGee helps Jim test the magnetometer. They bury a three-foot-long steel spike in a shallow hole and, without telling Terri the location, ask her to scan the area. A beachcombing metal detector Lou brought as a backup device has already failed.

Terri walks back and forth, the magnetometer on her back, staring at the little screen at her waist. At one point the device seems to register a slight signal, but the test is a bust. Terri says a plane engine would give off a much stronger response, but Jim walks away shaking his head.

With few remaining options and little time, Jim, Lou, Jetta, Jaana, and I walk along the rocks to the spot overlooking Balchen's X. Jim still hopes that the Duck is out there, and he again describes the flight path Pritchard might have taken. But Jaana's radar search of Essex Three, a heavily crevassed site closest to Balchen's X, turned up nothing, removing another once-promising point of interest. Out of earshot from the others, Jim concedes, "I'm starting to think we're not going to come up with anything."

Later that night, I find Lou resting his sore knees and still holding the gold flagpole ball.

"Don't get any ideas," I say.

"Don't worry," Lou says. "I'm not giving up."

With help from the safety team, he screws the ball into place.

AT A PLANNING session that night, Lou and Jim agree that there are no more reasons to search Essex One, Essex Two, Essex Three, or Points A, B, and K. Once the six highest-priority sites, now they're the glacial equivalent of dry wells. Tomorrow, Terri will carry the magnetometer to the lower-priority Point O, located on a slope roughly between Essex Two and Essex Three. Jaana's radar readings there were confounded by underground crevasses, so the magnetometer visit feels more like crossing an item off a to-do list than investigating a real prospect.

Despite the loss of time from the GPS mishaps, the 2012 Duck Hunt expedition will soon have cleared seven sites, one more than required by the Coast Guard's contract with North South Polar. It's getting colder on the glacier, and reports from Kulusuk Airport say a storm is coming our way. We have at most two or three days before we'll have to leave. With nothing better to do, Jim suggests that the radar team head to the farthest point yet, CRREL Point N, a mile beyond Essex Three.

Standing over a map in the command tent, Lou has an idea. He sees that the newly discovered Point N anomaly is near two other sets of coordinates, one given to him by JPAC, the Joint POW/MIA Accounting Command, and the other from the final crash report written about the PN9E in 1943. If Jaana is headed to Point N, Lou says, she should also survey the JPAC location and the 1943 point. He calls the historical point BW-1, because the crash report was written at the Army's Bluie West One base.

Lou admits that it's another long shot, probably the longest yet. Neither the JPAC point nor BW-1 registered as a hit on any previous aerial radar survey, which is why they weren't named among the priority sites. Also, long-standing doubts about the historic sightings of the Duck undermine confidence in the BW-1 coordinates. Still, there's nothing to lose but time.

MY DREAMS OF finding the Duck pretty well dashed, I head to the sleeping tents. On my way, John grabs my arm and invites me on a safety team adventure: a nighttime glacier hike. Joining us are Frank, Michelle, and Jaana, who despite all her radar work is up for more hiking.

Equipped with ice axes and roped together, we set off by moonlight toward Essex One, to find a crevasse that John noticed earlier and now wants to explore. We talk and laugh on the way, a momentary relief from what we all suspect is the expedition's looming failure. When we reach the crevasse, we lower each other one at a time into its crooked mouth. The opening leads to a cave filled with countless enormous icicles in translucent shades of blue, a secret underground spectacle that Michelle names "the chandelier room."

On our way back, undulating green curtains of northern lights stretch across the sky. Frank tells us to look away and then quickly look back. Each time, the shapes change, like wisps of luminescent smoke against a blue-black night. The sight gives me new appreciation for the misery of PN9E navigator Bill O'Hara. Anyone who wants to shoot the aurora borealis from the sky must know suffering beyond measure.

WITH SEARCH LOCATIONS dwindling, Jaana feels pressure to find something. Every day, upon returning to camp from a radar run, she hears a half-dozen versions of the same question: "Did you find anything?" Each time, she experiences the sadness that comes from her reply.

Before Jaana leaves with John and Frank to search the final three locations, Point N, JPAC, and BW-1, Steve aggravates her by asking if she'd be willing to run the radar not only around the three points of interest, but everywhere in between and on her way to and from the sites. Restraining her desire to tell him off, Jaana refuses, but the implied message plays on her nerves: we're desperate, so come back with something. Trying to stay cool, she tells herself, "I can't do more than cover as much as possible, and if I do not see anything, I do not."

Fortified by Michelle's breakfast egg burritos, Jaana, Frank, and John leave camp in their usual order: John out front to keep the lines straight when the radar work begins; Jaana in the middle with the gear; and Frank in the rear, watching the dragon tail— and Jaana—to keep both out of crevasses. They go first to BW-1 because it's the closest of the three sites to camp, just under 1.4 miles away. Using John's GPS, they find the coordinates and place a flag at the spot.

Expecting a repeat of the previous days' fruitless work, the trio begins walking one radar line after another to the southwest of the flag. Each line extends up to five hundred feet, to be certain the area is thoroughly covered. No luck. Next they move to the northeast side of the flag to run more radar lines. Partway through the second line, Jaana calls out, "Hey, John, can you stop?"

During four days of radar work together, the radar team has scanned nearly fifteen miles of glacier, not including the miles they've walked together back and forth from base camp and between the points of interest. Not once in that time has Jaana stopped in the middle of a line. But staring back at her from the little screen at her waist is something unusual. The glacier at BW-1 is almost free of crevasses, a near-solid block of ice some one hundred feet deep atop bedrock. A perfect ice cube of monstrous proportions. But now, Jaana sees a flaw in the cube.

Between thirty and forty feet below where they stand, the radar shows what Jaana calls "a large, clear anomaly." She settles herself and continues to work.

They start walking again, but when they reach the end of the radar line, Jaana surprises John and Frank again. Usually, they separate their lines by forty-five to fifty feet. This time, she asks John to lead her along a line close to the previous one. Neither John nor Frank asks why, but they know something's up. On the second pass, the anomaly announces itself again on the screen, a boomerang-shaped message from beneath the ice. Jaana asks her partners to place a second flag between the two lines, directly over the spot where she thinks the anomaly is located.

At 10:25 on Monday morning, August 27, the expedition's fifth day at Koge Bay, walkie-talkies set to the same frequency come alive: "Radar team to base camp," Frank says.

"Come in, radar team," Lou answers.

"We have a ten-meter anomaly at BW-1 position."

"Do you like it?"

A long pause ensues.

"She likes it. Over."

Everyone within earshot catches the significance of Frank's last comment. It's the first time that Jaana has been impressed enough by the sight and size of a hyperbola on her radar screen to alert base camp from the field.

Lou calls them back after warning me, "If I cry, don't take my picture."

He asks for Jaana, then says, "Is there anything different about this anomaly."

Jaana: "Yes."

Lou: "Can you please tell us what?"

Jaana knows what Lou wants to hear: it's the Duck. But she's a scientist, and she won't jump to conclusions. She says calmly, "This is in clear ice, with fewer crevasses." Jaana explains that the anomaly is large, and it's more than thirty feet deep in what otherwise appears to be solid ice. Also, it creates a radar response that goes all the way down to the bedrock, which makes it unlikely to be a crevasse. Yet until WeeGee melts some holes and drops the camera, there's no telling for certain what it might be.

Jaana's restraint notwithstanding, word of the BW-1 anomaly races through base camp. Terri and the magnetometer team are told to move from Point O to BW-1 as soon as possible. Jim calls Air Greenland to request a Hotsy airlift from Point A to BW-1, a 1.3-mile distance over crevassed terrain, made worse by a large area where there's a steep four-hundred-foot rise. The idea of the Hotsy team pushing it over the ice seems ludicrous and potentially dangerous. Lou goes as far as to say it would be impossible. Jim asks Air Greenland for fast service, but the first available helicopter won't arrive before late afternoon tomorrow. We'll take it, Jim says.

Lou swallows painkillers for his knees and scrambles to BW-1 to watch the magnetometer sweep. When Terri crosses the spot over the anomaly, her screen registers a reading "ten times higher than the ambient magnetic field." In other words, something metal appears to be buried in the ice. For the first time in days of walking atop the glacier, Terri has a hit. The magnetometer shows the same reading on several passes, but not all, leaving some doubt

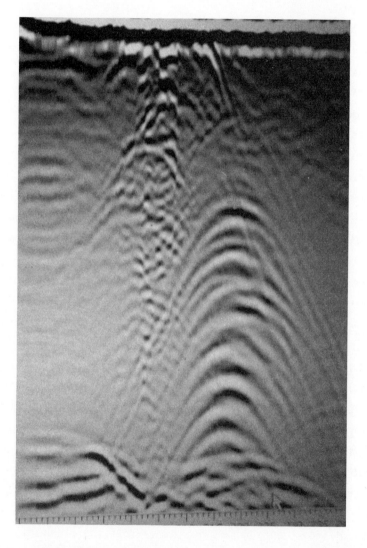

A SECTION OF THE RADAR COMPUTER SCREEN SHOWING THE TEN-METER ANOMALY AT BW-1. THE SMALL HYPERBOLAS NEAR THE SURFACE ARE ALMOST CERTAINLY A CREVASSE, WHILE THE LARGER ONES DEEP IN THE ICE RAISE HOPES AMONG THE DUCK HUNT TEAM. *(MITCHELL ZUCKOFF PHOTOGRAPH.)*

whether the machine is working properly. Still, Lou considers it confirmation of the radar finding.

"I said, 'John, Ben, Loren, give us a sign,'" he says. "And Terri started and stopped, started and stopped, and oh my God, it's there."

While Terri is at BW-1, the radar team moves to Point N and the JPAC point. They find nothing at either site. Like BW-1, Point N is almost solid ice, but with no anomalies worth noting. The JPAC site is so heavily crevassed that Jaana's screen fills with hyperbolas, making it almost impossible to pick out an anomaly if one's there.

After Terri's magnetometer hit at BW-1, Lou wants the radar team to return to repeat the survey. Jaana, John, and Frank run several lines in a new direction from the BW-1 flag, with the same positive results. Point N and JPAC are forgotten; all our bets are on BW-1.

BACK AT BASE camp, geophysics experts Jaana, Terri, and Bil are upbeat but restrained, knowing that the hyperbolas might be a crevasse, and the magnetometer's accuracy has been suspect. Still, Terri says that when she first heard about Frank's walkie-talkie call with Jaana's message from BW-1, she thought, If she called it out, it has to be something significant. Jaana wears a poker face, but privately admits feeling "full of energy, really happy and excited."

In the command tent, Jim is optimistic but cautious. For one thing, if BW-1 is the crash site, he has to abandon his hard-earned theory about Balchen's X, which is about a mile largely downhill from BW-1. Glaciers don't move uphill, so even with glacial movement, there's no chance that the Duck migrated from Balchen's X to BW-1. In other words, if BW-1 is the Duck's resting place, Balchen's X was in the right general vicinity, but misplaced.

As NIGHT FALLS, Lou and I agree that we might never have a better opportunity to break out Shackleton's Scotch. If we wait and the BW-1 anomaly is a bust, we'd be drinking fine whiskey to wash away the bitter taste. If, on the other hand, the news is good, opening the historic blend will mark the start of our celebrations.

I pull the bottle's wooden case from my duffel bag as everyone gathers in the dome. Our plastic cups held high, Lou offers a toast: "To Lieutenant John Pritchard, Radioman First Class Benjamin Bottoms, and Corporal Loren Howarth. Your families want you home. We're here to bring you home, and may we be successful." His cheeks flushed, his silver hair flowing from under a North South Polar baseball hat, Lou thanks us one by one for our contributions to the mission.

I've never seen him in finer form or the team in better spirits.

The question now is whether, as on Shackleton's failed mission to the South Pole, the best part of our expedition will be the Scotch.

23

"SOME PLAN IN THIS WORLD"

MARCH–APRIL 1943

CAUTIOUSLY AVOIDING CREVASSES during the first mile to the Motorsled Camp, Monteverde, Best, and Spina each walked under his own power, as did the three-man trail team. The nine dogs pulled a main sled, behind which was a tow sled loaded with everything the men and beasts would need until Barney Dunlop's Dumbo returned to fetch them or Pappy Turner's B-17 resupplied them.

The team's lead dog was Rinsky, a fierce husky born in Antarctica and brought to Greenland by Healey, its owner. Two other dogs on the team were called Pat and Mopey. Raised in barren lands with no trees or hydrants, male sled dogs had no targets upon which to relieve themselves, so they usually squatted rather than raising their legs. Sometimes, though, a man's pant leg might get watered in a display of disdain or dominance. Greenland dogs tended to be aggressive, often fighting among themselves for scraps and power. Straddling the line between wild and tame, most had little use for affection or human company. Some were whip-smart and some were dumb as sleds. Some were handsome and some were not. All were tough and seemingly immune to pain and cold. Most seldom barked, but they'd howl like their ancestors at night and at meals. When tired, they'd curl into tight balls of fur, their faces against their flanks, to sleep through Arctic winds.

As the men and dogs marched across the snow-covered ice, Spina was the first to falter. During more than four months since the crash, the farthest he'd walked was fifty feet in pursuit of the milk can. He tried to keep up but soon he fell to his knees every twenty or thirty feet. He'd rise and stumble forward with his eyes shut, then fall again into the snow. After one fall Spina made no effort to rise. He felt resigned to die in place. Strong had other ideas; he bundled Spina aboard the tow sled. Best fell next. Monteverde teetered, tempted to pitch forward into a snowbank and sleep forever. Strong called a halt.

Dolleman raised a tent and climbed inside with Best and Monteverde. They'd remain behind to rest while the others raced ahead to the Motorsled Camp. Strong and Healey continued on foot while Spina reclined on the tow sled. As they hustled across the ice, Strong stepped over the edge of a crevasse. But he was no greenhorn in Greenland; he held tight to a rope attached to the dogsled, and the sled's momentum pulled him up to safety.

Along the route, they planted red warning flags to mark crevasses and yellow guide flags to mark the safe path. When they reached the empty Motorsled Camp, Strong and Spina climbed into a tent. Healey and the dogs swung around to retrieve Dolleman, Monteverde, and Best.

Once reunited, the six men spent the next two nights in tents on the ice. During the day, Strong, Dolleman, and Healey enlarged and improved the warren of snow caves left behind by Don Tetley, Harry Spencer, and Bill O'Hara. When they climbed inside, Monteverde, Spina, and Best were astonished: their friends had created an underground ice palace.

The entrance was a large hole with a fifteen-foot staircase cut from snow. That led to a hallway about six feet wide, twenty feet long, and ten feet high, with an oil stove at the far end to keep the lair warm and to dry their clothes. Along the hallway were openings that led into small sleeping rooms, like berths on a train. Each was about five feet off the ice floor, to keep water from accumulating in them. The hallway also led to a kitchen with shelves cut into the ice and a vent to the surface for cooking and heating fumes. A large room off the kitchen was the pantry. Past the stove at the end of the hallway was a second set of stairs, leading down another ten feet, to a latrine carved from ice and snow. The three remaining PN9E survivors were so impressed that they renamed the Motorsled Camp: now it was the Imperial Hotel.

The six men enjoyed several days of good weather, during which Turner's B-17 boosted their supply cache. But several days of storms followed. The wind was so strong and the snow so fierce that Healey brought the dogs down into the human quarters. One husky that refused paid for his disobedience with a case of frostbite, though he recovered. The dogs treated the underground maze like a kennel, fighting and running through the rooms. When the dogs settled down, they became warming blankets for the men, who tucked their sleeping bags against them at night.

TWO WEEKS PASSED during which Strong, Healey, and Dolleman cared for the needs of their three Imperial Hotel guests. Healey cooked, and with a wide variety of available supplies he took dinner orders from each man. Healey didn't like coffee, so he resisted making it, but he kept a pot of tea boiling on the stove around the clock. They stayed up late every night, talking by candlelight and telling jokes. Spina, the jokester of the PN9E, credited Dolleman for keeping them all in stitches. Between laughs, the three trail men told the three fliers stories from their Arctic adventures.

When it was light they climbed up from the cave, and Healey

strapped Spina to the sled for daily exercise runs with the dogs. They made multiple passes over the designated landing area, to tamp down new snow.

When Strong was still at Bluie East Two, Don Tetley gave him a detailed map of the Motorsled Camp that included the general location of the buried motorsled. When the weather cleared, Strong decided to get some exercise by digging for it. Dolleman and Healey joined in, and in time so did Monteverde and Best. Spina, his arm still recovering, appointed himself foreman. They dug for three solid days before finding the missing motorsled under twenty feet of snow.

AT BLUIE EAST TWO, rain replaced snow and coated the runway with slush. A bigger worry for Balchen was that melting would make the snow at the Motorsled Camp/Imperial Hotel sticky, preventing the PBY from taking off after it collected the six men and dogs. Adding to Balchen's concerns were high winds that wreaked havoc on the two PBYs on the tarmac. Both suffered broken ailerons, the hinged sections on the trailing edge of the wings that allow an aircraft to bank left or right. Time was passing, and repairs added to the delays.

ON APRIL 5, 1943, nineteen days after the trail team arrived on the ice, Harold Strong radioed Bluie East Two with good news. The ground temperature was relatively warm and the wind had taken the day off. Balchen ordered him to break down the sled and get ready to load everything aboard the rescue plane.

Worried about the Dumbo's weight on takeoff from the ice, Balchen had crews strip the plane of everything not essential to flight or stability. He filled its fuel tanks with only enough for the round-trip, plus a little extra for safety in case of delays.

Before the Dumbo arrived, Pappy Turner flew over the Imperial Hotel in his B-17. Spina got on the walkie-talkie and promised his friend Carl Brehme, Turner's engineer, that he'd buy him the biggest steak he could eat if the rescue succeeded. During the same conversation, Monteverde heard good news: he'd been promoted to captain during his nearly five months on ice.

As the Dumbo approached, the six men on the ground turned themselves into human weathervanes: they lined up facing into the wind, a signal they'd devised to let pilot Barney Dunlop know the wind direction. He landed as smoothly as he'd done twice before, sending plumes of snow into the air that briefly obscured the plane. Dunlop taxied in a wide circle and stopped near the waiting men.

Balchen wanted to hustle the six men, dogs, and equipment aboard to keep the Dumbo's belly from freezing to the ice. But the plane's crew hopped off, wanting to take pictures and to welcome the long-missing men. When the greetings ended, the dogs

STANDING ATOP THE IMPERIAL HOTEL ARE (FROM LEFT) BARNEY DUNLOP, HENDRIK "DUTCH" DOLLEMAN, ARMAND MONTEVERDE, JOSEPH HEALEY, AND PAUL SPINA. *(U.S. ARMY PHOTOGRAPH BY BERNT BALCHEN.)*

and equipment were loaded first, followed by the six men. Balchen noticed that Monteverde, Spina, and Best boarded in silence, as though unable to believe that their long wait was nearly over.

But once again, the Dumbo's belly froze to the ice. Strong and Healey got out and rocked one wing, while Dolleman and the plane's engineer, Alex Sabo, rocked the other. Dunlop gunned the engines and the plane shook as it fought to free itself. When the Dumbo broke loose, Dunlop made four laps around the field. Each lap allowed another man on the icebreaking crew to board on the run through the side blister.

Dunlop steered the plane into position for a long, uphill takeoff run. The passengers stood as far back in the tail as possible, to make it easier for the Dumbo's nose to lift. Dunlop leaned hard on the throttles and the plane sped across the ice, rising three or four feet then dropping back down, then rising, then dropping again. Dunlop turned around and tried a downhill run, but again he couldn't gain enough lift. He eased back on the throttles just short of a crevasse.

The Dumbo had four pilots aboard: Dunlop, Balchen, Monteverde, and the copilot, Nathan Waters. By the second failed attempt, none could have doubted that they faced a nasty combination of too much weight from the passengers, dogs, and cargo, and too much friction from the slush coating the Dumbo's fuselage. A strong headwind would have lowered the ground speed needed for takeoff by creating greater airflow over the wings. But even that might not have been enough, and it was a moot point, anyway. On the one morning when they needed it, the notorious Greenland wind was nowhere to be found.

The Dumbo's twin engines began to overheat, but Dunlop pressed on. He turned the plane uphill for one more run. As he gained speed, the right engine burst into flames. A line broke and shot black oil onto the wings and fuselage. Dunlop shut down both engines, and crew members jumped out to extinguish the fire. The vacuum pumps were damaged, the fuel pressure gauge line was burned, the cowling on the damaged engine was melted, the engine had lost oil, the exhaust rings were burned out, and the entire works were black with soot. The engine wasn't quite dead, but any hope of flying that day was gone. Dunlop taxied back to the Imperial Hotel.

Monteverde, Best, and Spina couldn't believe their bad luck. They seemed doomed to remain on the glacier.

But Balchen and Dunlop weren't done. The crew began emergency repairs, holding together the damaged, twelve-hundred-horsepower engine with steel straps from equipment cases. Pappy Turner dropped them fresh oil and a replacement oil line, but installing it would mean the time-consuming task of taking apart the engine. The repair crew held off, waiting for a decision from Balchen on whether to make the repairs quick and dirty, or take several days to install the new oil line. His decision depended in part on when he thought the winds would return. They had no way to anchor the Dumbo, so a powerful storm might toss the plane like a dry leaf.

Everyone left the plane and holed up in the overgrown snow cave, discussing their options and dining on a meal prepared by Balchen himself. Balchen thought aloud about the challenges ahead. The second PBY was damaged and unavailable, so they had neither a backup plane nor a backup plan. Getting the three PN9E survivors off the ice soon meant relying on the Dumbo parked outside the Imperial Hotel.

Balchen worried that installing the new oil line would take too long and endanger the mission. The plane would be exposed to windstorms without being tied down. Or, it might freeze to the ice so solidly that no amount of rocking would free it. Balchen believed that they had one last chance to get airborne without the new oil line. To do so, Barney Dunlop could use both engines for takeoff, and then rely on the undamaged engine for the return flight to Bluie East Two. Balchen calculated that just enough fuel remained if they lightened the load. Having made his decision, Balchen announced that he, Strong, Healey, and Dolleman would remain behind, along with their dogs, camping equipment, supplies, and sled.

The plan was pure Balchen. Taking off from the ice cap was, by itself, a dangerous maneuver in the best conditions, accomplished only four times previously, twice by Pritchard in the Duck and twice by Dunlop in the Dumbo. Now, ignoring the possibility of an explosion or a crash, Balchen wanted Dunlop to take off with a half-blown engine, and then fly the survivors back to the base with barely enough fuel to get there. Meanwhile, Balchen and the trail team would trek to the coast, with little hope of rescue if disaster struck.

The three PN9E survivors were skeptical, but they trusted Balchen. And after five months on the ice, they were ready to try almost anything.

After dinner, Balchen and Dunlop spent the night in the cave with the six regular guests of the Imperial Hotel, while the rest of the crew returned to the plane. Monteverde, Best, and Spina went to sleep praying for good luck and good weather.

THE SURVIVORS' PRAYERS were answered almost too well. The day was clear and the air dead calm the morning of April 6, 1943. Balchen wanted at least some headwind to lift the Dumbo, so they ate breakfast and waited. They also waited for Pappy Turner's B-17 to arrive overhead, so he could follow the Dumbo back to the base. With Turner tracking the plane, rescuers would know where to look if its engines failed.

When winds picked up in the afternoon, Balchen and Dunlop decided it was time. The survivors and the Dumbo crew scrambled on board while Balchen, Strong, Dolleman, and Healey rocked the plane to break its icy cradle.

Dunlop raced down the glacier runway. He tried for nearly a mile, occasionally bouncing several feet off the ground. But just as on the previous day, he couldn't gain lift. He turned the Dumbo around and tried in the opposite direction, but again had no luck. Their chances were running out. The plane's already slim margin of fuel was nearly gone. The damaged engine wouldn't take much more strain. Dunlop got out and strategized with Balchen.

"If I hadn't flown in this ship before," Dunlop told the legendary pilot, "I'd almost say that this ship wasn't built for flying."

Confident as ever, Balchen claimed that all they needed was more wind. The pilots and crew stood around, talking and taking pictures. After several hours stronger winds arrived. One cost of the delay was the absence of an escort: Turner's B-17, running low on fuel, had returned to Bluie East Two.

Balchen and the three trail men shook hands with the men boarding the Dumbo, as both groups headed toward uncertain futures.

THE SLED DOG TEAM PASSES THE PBY DUMBO AS IT WARMS ITS ENGINES. *(U.S. ARMY PHOTOGRAPH.)*

THE FOUR-MAN TEAM of Balchen, Strong, Dolleman, and Healey soon began the demanding march with their dogs to Beach Head Station. They followed a winding route that stretched more than forty miles with detours to avoid crevassed areas.

Later, Balchen recounted the journey with undisguised pride: "I have no instruments along for land navigation, and I have to guide the party by dead reckoning. With a prismatic pocket compass and a protractor, I make computations in pencil on a diary page, and clock off our mileage on a distance-measuring wheel fastened to the runner of the sled. For five days, we hole up [during] a *williwaw*"—a frightful storm—"staking our dogs securely, and digging under the snow ourselves to ride out winds up to 150 miles an hour. Ten days later, we have worked our way to the coast, through drifts and *sastrugi* as high as three feet."

All four men and their dogs arrived safely at Beach Head Station. On April 18, 1943, a plane picked them up there and returned them to Bluie East Two.

AFTER BALCHEN AND the trail team rocked the Dumbo free from the ice, Dunlop set off down the glacier runway once more, heading uphill into the resurgent wind. Again the Dumbo bounced along the way, but this time the wind rushing over the outstretched wings provided enough lift to separate the plane from the ice cap.

With the damaged engine straining, Dunlop gained altitude slowly. Low on fuel and losing oil, he cut the dying engine when the Dumbo was about one thousand feet above the ground. Then he adjusted the angle of the right propeller so it sliced through the wind, a technique called "feathering the engine." Otherwise, drag created by the shut-down engine would have threatened the plane's ability to fly. But Dunlop's prudent move triggered a new problem.

Without power from the right engine, the Dumbo hung in the air as though planning a return to earth. Balchen's scheme seemed to be failing, and the passengers feared for their lives. After a heart-stopping pause, the plane went into a dive. Dunlop fought in the cockpit, but the Dumbo kept losing altitude. He leveled off, but soon the plane was barely more than fifty feet above the ground.

Spina looked out the window in the Dumbo's side blister and saw the ice cap rushing toward them. He and Best thought they were about to crash, ending their ordeal and their lives in the cruelest possible way. Harold Larson, the Dumbo's radioman, saw the survivors' white-knuckled distress. He patted them on the backs and told them everything would be fine, whether he believed it or not.

No one was comforting anyone in the cockpit. Fearing disaster, Dunlop resolved to restart the damaged engine to gain altitude. It was a risky move—if the engine exploded, they'd be done for. But if they didn't want to crash, they needed its power, even briefly.

Dunlop restored power to the engine; it strained and complained, but it held together. With both propellers spinning, Dunlop pulled back hard and got the Dumbo's nose pointing upward. But now he faced a new obstacle: the mountains that guarded Greenland's coastline. Dunlop demanded that the plane continue to climb, hoping to gain enough altitude while both engines worked to clear the highest peaks. In the cabin, the passengers watched silently through the windows as the mountains approached.

As Dunlop struggled, the instrument panel before him displayed a terrifying sight: the cylinder head temperature gauge for the damaged right engine was deep in the red, far past the danger zone. Leaking oil and pushed beyond its limit, it threatened to catch fire and explode. With no other choice, Dunlop continued to climb.

The Dumbo rose, and the jagged mountains passed beneath

them. When Dunlop felt confident that they'd cleared the peaks, he shut down the damaged engine and began to lose altitude. He pointed the plane east toward the water, in case they had to employ the seaplane's buoyant qualities. That move reflected Dunlop's newest worry: using the second engine for part of the flight hadn't been part of Balchen's calculations. Now they were nearly out of fuel.

Listening to the crew's conversations over a headset, Monteverde heard Dunlop ask the flight engineer how much fuel remained. The disturbing answer was, about 120 gallons. The Dumbo was burning seventy-three gallons of fuel an hour, and was still more than an hour from Bluie East Two. Monteverde could do the math. Based on their distance and fuel consumption rate, they'd have perhaps thirty gallons to spare. For a plane with a fuel capacity of almost 1,500 gallons, that translated as a teacup of water to sustain a thirsty elephant. A strong headwind could suck the Dumbo dry.

FOR THE NEXT hour, Dunlop tried every trick in his pilot's bag. Yet nothing could prevent the plane from losing altitude. From his desk outside the cockpit, radioman Harold Larson called Bluie East Two to say that the Dumbo might not make it. Without Pappy Turner's B-17 as an escort, no one would know where it went down. The base answered that it would send a new spotter plane, and soon a twin-engine Beechcraft AT-7 appeared alongside the Dumbo.

Somehow, Dunlop kept the air beneath his wings. About five minutes from the base, he spotted the mouth of the fjord leading to Bluie East Two. Dunlop asked again about fuel. The engineer told him that the gauges read empty.

Dunlop faced a choice upon which all their lives depended. He could aim toward the fjord for a controlled water landing, or he could go for broke. Dunlop chose the latter. He'd fly toward the base until the last fumes were gone and the lone engine quit.

Then he'd glide the rest of the way and make what pilots call a dead-stick landing.

Dunlop told everyone aboard the Dumbo that if they heard the remaining engine sputter out, they should prepare for a crash. The men in the cabin piled bags and equipment against the hard metal bulkheads to soften the blow. They sat together with their backs against the cushioning material and braced for impact.

As the Dumbo lost altitude, Dunlop lined up for his one shot at an approach. Now he discovered yet another problem. The dead right engine controlled the hydraulic pump for the landing gear. Using a hand crank, his crew lowered the main wheels manually, but they had no manual control for the nose wheel. Dunlop would have to bring the Dumbo down onto a hard runway on its snout.

This was the very danger the plane's manufacturer had expressed two months earlier in the message to Balchen. Too much pressure on landing might collapse the forward bulkhead. But with no fuel and a dead engine, Dunlop had no choice.

He touched down with the plane's main wheels. Dunlop fought to keep the nose up and away from the landing strip, without raising it too high and bouncing the tail on the ground. Within seconds, he began to run out of runway. Dunlop eased down the Dumbo's nose, to use it as a brake. The maneuver had to be handled just right—fast enough to stop, but not so fast and hard that he'd break the plane in two.

When the nose touched down, the Dumbo made a sharp turn, veering directly toward Pappy Turner's B-17, which happened to be parked at the far end of the runway. Dunlop braked as hard as he could. The men in the Dumbo braced for impact. Turner's bomber loomed ahead. The two planes, one in motion, one a stationary target, both of them committed to saving the PN9E crew, closed to fifty feet.

Then forty. Thirty.

Fifteen feet from Turner's B-17, the Dumbo skidded to a stop.

The plane carrying the last three survivors of the wrecked PN9E balanced motionless on its nose with its rear end high. The Dumbo looked as though it was taking a deep, well-deserved bow.

THE BASE AT Bluie East Two emptied to greet them. The survivors grabbed handfuls of dirt from the runway to celebrate their deliverance from ice cap purgatory. If they had continued their matchstick calendar to the last day, the top row would have displayed four sticks, the bottom row six. They'd been on the ice from November 9, 1942, to April 6, 1943. One hundred and forty-eight days.

Armand Monteverde, Alfred "Clint" Best, and Paul Spina joined Lloyd Puryear, Alexander Tucciarone, Harry Spencer, and William "Bill" O'Hara as PN9E crash survivors rescued from the glacier beyond the Koge Bay fjord. The five C-53 crewmen they'd hoped to find remained lost. Also left behind were PN9E fliers Loren "Lolly" Howarth and Clarence Wedel, and would-be rescuers John Pritchard Jr., Benjamin Bottoms, and Max Demorest.

SOME FORTY-FIVE YEARS later, a fully recovered Clint Best told his grandchildren about the crash, the bitter cold, the deprivation, and the fear. He explained how the hardships were offset by the caring he'd felt from his friends. Best left out the worst parts, so as not to upset the children, but he did mention his bouts of dementia.

Above all, Best marveled at the efforts made to save him and his crewmates. "Money was never an object. If there was someone out there that needed rescuing, the air force went out to the rescue," Best told his family. "They never give up. They never gave up on us."

Despite the passage of time and the wisdom of age, Best struggled to describe how it felt to have survived. "As I've gone on through the years, it's hard to figure out how you can be with eight other people, and one falls into a crevasse and disappears. It's, 'Why me, Lord?' In this case it's, 'Why him?' And the radioman goes after a plane and he gets killed, and the lieutenant comes down on the motorsled to rescue you, and he gets killed. You wonder why it's all of them."

Finally, Best settled on divine intervention: "I figured God must have had some plan in this world, that He let me along with the others remain."

24

DOWN TO THE WIRE

AUGUST 2012

THE DOME TENT is alive with excitement about the anomaly at BW-1 as we crowd together for breakfast on Tuesday, August 28. Between sips of coffee, Jaana happily displays a radar image on her laptop. She explains that it shows small hyperbolas near the glacier surface that are unmistakably a crevasse, and bigger and more dramatic hyperbolas deep in the ice that suggest a large UFO: an unidentified frozen object. The mood is light, and Alberto gets laughs by teasing Bil about the single "l" in his name: "What's Bil short for, Bill?"

But enthusiasm leaches from the dome when Steve announces that a half-dozen team members must return to Kulusuk aboard the helicopter that's due later today to airlift the Hotsy to BW-1. The decision to begin "demobilizing" from the glacier comes jointly from the Coast Guard and North South Polar, and takes the form of a plan agreed to by Jim, Rob, Lou, and Steve. Although the weather remains cold and clear at Koge Bay, a major storm is headed our way. Multiple helicopter trips will be required to get everyone and our equipment back to Kulusuk. If flights are canceled, some or all of us might be trapped here in dangerous conditions.

Bil and Alberto volunteer to leave to avoid canceling other obligations, but no one else wants to miss the hole-melting, camera-dropping finale. Steve clears his throat and reveals the names of the soon-to-depart: "So, going out today are myself, Terri, Ryan, Alberto, Michelle, and Bil. OK? Everybody else will be remaining, and the remaining group will figure out the lifts for tomorrow and Thursday."

Several people object on Michelle's behalf, knowing how much she wants to remain in camp. Steve wavers, asking Jim if he knows which Air Greenland helicopter is coming and whether we might fill it with more equipment and fewer people. But Jim wants to stick to the plan they'd made before entering the dome.

"We need to get people off the ice. That has to happen," he says.

Steve finds a way to commiserate with the rank-and-file yet also support the team leaders' decision: "Every one of us wants to stay. Unfortunately, six of us have to go back."

Steve's mixed message and his inquiry to Jim about fewer people leaving create an opening, and Michelle gains more voices of support. Frank, however, is silent, not wanting his relationship with Michelle to be seen as coloring his judgment as safety leader. Faced with more rumblings on Michelle's behalf, Steve throws up his hands. "I'm not making a command decision on this one."

That sets off Rob, who as second-in-command of the Coast Guard contingent occupies a parallel position to Steve's on North South Polar.

"You're the command," Rob tells Steve. "That's the job. You want me to make it?"

The tent goes silent until Bil tells Steve what everyone is thinking: "The gauntlet's thrown on the ground there, bud."

Lou steps into the fray, again raising the idea of sending gear before people. Discussion moves to the agenda for the day then circles back to the helicopter. Ultimately, Lou expresses support for the original plan, and both he and Jim say they believe that the six people Steve named should be the first to leave.

Steve seems distracted, agitated by the public conflict with Rob about command. He detours the discussion to describe how the campsite should be broken down and to list tasks needing completion before the helicopter arrives. Finally he musters himself to declare who stays and who goes.

"Michelle," Steve says, "I'm sorry to be the bearer of bad news, but—"

"I'm useful," she tells him, listing her skills in mountaineering, glacier climbing, medical care, and elsewhere.

"Well, you know what," Steve says, reversing field, "I'm going to make the command decision. Rob, you're going."

Blindsided by Steve's switch and frustrated by a week of leadership he finds lacking, Rob can't restrain himself. "See how that felt? Feel good?" he asks Steve mockingly. "That's what it feels like—command!"

"Oh, really, thank you," Steve shoots back. "Thanks for telling me what command feels like. Ever have a combat command?"

"You were saying you didn't want to make a command decision," Rob answers. "That's your job."

The glow is off the dome, and tension is now the order of the day.

BEFORE THE MORNING meeting ends, the six departing team members are told to gather their belongings and break down their tents. With no radar or magnetometer teams going out, there's not much for the rest of us to do before the helicopter arrives. Eager to get as close as possible to the anomaly, Lou, John, Frank, Michelle, Jaana, and I head across the glacier to BW-1. John carries an ice auger to drill holes to see if we might find pieces of the Duck near the surface. We also bring the beach-combing metal detector, more for kicks than with any expectation that it would be powerful enough to find the anomaly so deep in the ice.

It's an hour-long uphill climb, and we're winded when we reach the orange flag. With time to spare, we soak up the sun and a spectacular view of Koge Bay, clogged with icebergs as big as cargo ships. It's the clearest sightline we've had all week of the glacier where the PN9E crashed, a sharply rising, impossible-to-see hazard for any pilot "flying in milk." Jaana and I take turns passing over the anomaly with the metal detector, which buzzes only when we accidentally bang it against our legs. John and Frank use the auger to drill enough holes to make a coffin-sized opening in the ice, but there's no sign of the Duck. Soon the hole fills with frigid water flowing through the glacier, putting an end to the drilling.

The sun is warm and skies are blue, so several of us use our coats as blankets and stretch out for glacier naps. By noon, Michelle leaves to make lunch and Lou goes with her to oversee the first stage of base camp breakdown. A tempest of problems awaits them.

AFTER SPEAKING WITH airport officials by satellite phone, Steve reports that a thick fog has grounded the helicopters at Kulusuk at least until tomorrow afternoon. When the sky clears, Air Greenland will send a helicopter to move the Hotsy and to get as many people as possible off the ice ahead of the storm. In a small way, it's good news for everyone who didn't want to leave the glacier today. But that's little consolation compared to the larger costs.

No helicopter today means no Hotsy move tonight. That means no hole-melting tonight or tomorrow morning to explore the anomaly. Even if the helicopter does come late tomorrow and moves the Hotsy, the storm might leave us no time to investigate BW-1. In that case, our only option would be a profoundly disappointing backup plan: place a satellite-tracking device above the anomaly and go home. We wouldn't know whether we'd found the Duck's crash site and the resting places of Pritchard, Bottoms, and Howarth. A year or more might pass before some or all of us could return. That is, assuming enough money might materialize to support yet another expedition. With no evidence more solid than radar and magnetometer hits, it's doubtful. The thought casts a pall over camp.

In hushed conversations held in clusters around the rocks and tents, there's talk of riding out the storm and moving the Hotsy after the bad weather passes. But that idea is soon squashed. We don't know how bad the storm might be or how long it might last. And even if we did hunker down, we'd have little or no time afterward to melt holes. The Coast Guard's C-130 is due to return to Kulusuk in three days, and that's the only way to get our four tons of equipment back to the United States. There's no hope of delaying the big plane, and the cost of flying the gear and everyone on commercial airlines is beyond prohibitive.

The bottom line is that if we want to investigate the BW-1 anomaly, it's now or possibly never. Without a helicopter's help, we have about twenty-four hours to somehow move the Hotsy 1.3 miles from Point A to BW-1, largely uphill and across innumerable hazards.

Lou, Jim, and several others stand in a tense knot on the ice field, discussing and rejecting one option after another. Lou is already on record as being uncharacteristically pessimistic about our chances: "There's no way we can move the Hotsy over land." It's physically impossible, he thinks, and there's a danger that a bridged-over crevasse might give way and swallow some or all of us.

Writing in his journal, Nick sums up the risks by recording

observations from his first trip to the anomaly site: "The route to BW-1 was an indirect path through a series of open crevasses, surface meltwater channels, and hidden moulins [deep vertical shafts within a glacier]. These last ones raised my eyebrows. Not far beneath the surface you could hear water running. Not crevasses, but drainage tunnels, the plumbing of the glacier. Fall into one roped and you had a chance of getting out. Fall in unroped and you might just get flushed down to the fjord over the next thousand years."

Jim isn't cavalier about safety, and he doesn't pretend to be a glacier expert, but he refuses to surrender. Just as Lou has scrounged and sacrificed to be here, Jim has put his reputation on the line. With the anomaly at BW-1 staring at him from Jaana's screen—he's a frequent customer asking for a look—Jim refuses to return to his desk at Coast Guard Headquarters without knowing what's down there. Despite concerns from Nick and others, Jim believes that the safety team can find a solid path to move the Hotsy to BW-1. To make the trip less arduous, Jim and several others wonder whether the lids of large Pelican cases might be converted for use as sleds under the Hotsy's wheels.

Several team members discuss abandoning the Hotsy altogether and carrying a second auger, fuel, and a pickax to BW-1, to see if those tools might reach the anomaly. The idea of leaving ahead of the storm, without any melting or drilling at BW-1, also hasn't been ruled out.

As the chief hole-melter and Hotsy wrangler, and someone with a seemingly inexhaustible appetite for hard work, WeeGee keeps tabs on the swirling discussions. He moves among the groups, listening more than talking, assessing the ideas and attitudes of the would-be planners. After suffering through hours of inactivity and indecision, WeeGee's frustration gets the best of him. He has no intention of leaving before he can perform exploratory surgery on the glacier. He's certain that the Hotsy is our only hope, and he dismisses the idea that Pelican lids might help us push the seven-hundred-plus-pound machine to BW-1.

WeeGee disappears from the brainstorming sessions and grabs the second aluminum extension ladder bought in Keflavík. He separates the ladder into two parts, each twelve feet long, and carries one to the rocks past the toilet tent. He finds two closely set boulders and jams the ladder's end into the opening. What seems like the act of a frustrated madman reveals itself as the inspired work of an innovator. Pulling down on the ladder, he bends it at the second rung. He pulls it out, inspects it, then bends it some more. When the ladder's end is curved upward like the tip of a ski, he flips it around and does the same to the other end.

His orange boots stomping against the ice, WeeGee carries the custom double-curved ladder toward Jim, Lou, and the other planners. He halts twenty yards away and wordlessly slides it in their direction. The ladder skims across the ice like a sharpened skate and stops near their feet.

Jim gets the idea immediately. "Hell, yes," he says. "Money."

Bil wraps WeeGee in a hug: "You gave nobody any choice. This is how we're doing it."

WeeGee repeats the bending process with the second half of the ladder, and soon the plan is apparent to everyone. With the curved ladders serving as strong, lightweight runners, WeeGee intends to turn the Hotsy into a giant sled and use us as huskies.

Lou snaps back to his natural optimism and muses about moving the Hotsy tonight. Nick and several others tell him he's nuts—the glacier's surface is too slushy from the day's sunshine. He relents only when WeeGee declares that we'll wait until morning, when the route to BW-1 will have frozen overnight. WeeGee's primary concern is about someone getting hurt, but he also worries that the ungainly Hotsy might tip over and break if conditions aren't close to ideal.

As dusk approaches, WeeGee, Jetta, and Nick leave base camp

for Point A to fasten the twin ladder skis side by side under the Hotsy. En route, WeeGee explains that he bent both ends of each ladder as a precaution, so the Hotsy can be pushed in either direction if one ladder end breaks or nose-dives into a ditch. When he says "ditch," most of us hear "crevasse," in which case a double-curved ladder won't do anyone much good. On the other hand, even after WeeGee's end-bending trick, about nine feet of each ladder makes contact with the glacier, or enough to span the widest crevasses we anticipate. As a last-minute tweak, WeeGee places shovels underneath two of the Hotsy's tires, so they sit more squarely on the ladders' rails.

Fully assembled, with the twin ladders strapped to the upper frame suggesting biplane wings, and the turned-up ladder skis evoking a central pontoon, the Hotsy's homage to the lost Duck is complete.

"Good morning, campers!"

It's 5:15 a.m. on Wednesday, August 29. A wide-awake Wee-Gee marches among the tents sounding reveille. The rest of us crawl bleary-eyed from our sleeping bags out onto the ice. The air is 23 degrees Fahrenheit, but swirling winds make it feel closer to zero. Icicles form instantly when Michelle pours water from a jerry can to prepare breakfast. There's little chattering except our teeth as we gather in the dome to eat. Everyone knows what's riding on today, and thoughts bounce from safety to our collective strength then back to safety. Jim breaks the silence by predicting that it will take four unrelenting hours to move the Hotsy to BW-1, a grind of fewer than six hundred yards per hour.

In groups of twos and threes, we trek to Point A, and by 6:20 we assume our positions. It's an all-hands operation, with nine of us on the pushing ladders and nearly everyone else either yoked to a harness attached to the front of the Hotsy or walking out front to scan for hidden crevasses. A half-dozen orange flags bundled together on the Hotsy give it a festive look, as though it's a crude carnival ride being moved onto a frozen fairground.

During the early going, Frank acts as the last line of defense, holding a rope tied to the rear of the Hotsy in case we lose control and it slides left, right, or headlong down the glacier. I doubt that Frank could stop it, but having him there is comforting. Considering his strength and resourcefulness, I suspect that he'd manage something. When everyone's in place, Ryan calls out, "On Prancer! On Dancer! On Comet! On Vixen!" Doc Harman adds a benediction: "We're going out for Pritchard because he would have gone out for us."

The first two hundred yards are smooth, and under our power the Hotsy ladder sled fairly glides across the glacier. We hit a rough patch of ice and bounce across it. I'm on the right side of the front pushing ladder. WeeGee is on the left side. He catches my eye across the machine and smiles. We both know what he's thinking.

We pick up the pace as we approach the first bridged-over crevasse, hoping to gain enough momentum to fly across without testing its load-bearing capacity. Increasing speed also allows us to enjoy the pleasant illusion that, like barefoot walkers on hot coals, the faster we cross the less likely we'll get burned. Grunts, groans, and shouts of "Push!" ring out. The ladders shudder but hold and we clear the crevasse.

On the other side we rest, congratulating WeeGee on his invention and ourselves on our teamwork. Conflicts and tensions of the previous week fade away, replaced by fatigue and the shared goals of reaching BW-1 and firing up this awkward beast on improvised skis.

WE PUSH ONWARD, crossing smaller crevasses and shallow channels where meltwater drains toward the fjord. We rearrange the crew for maximum power as we approach the steep four-hundred-foot ice-covered hill that we've known from the start will be the true test. Out front on the pulling ropes are Alberto, Nick, and Rob on the left side, and John, Jaana, and Frank on the right. On the front pushing ladder, left to right, are WeeGee, Jim, me, and Ryan. On the back ladder are Terri, Doc, Michelle, Lou, and Bil. Steve alternates between pulling a tow rope and shouldering a heavy ice drill while navigating our path, and Jetta helps everywhere she can while exhorting us and photographing the work.

The Hotsy's weight seems to double as we begin the climb. Muscles strain, joints ache, faces contort. Breathing grows louder. Joking disappears. Maybe Lou was right and this is impossible.

After a hundred yards we rest and drink from canteens until heart rates drop and energy rises. Halfway up the hill we encounter the biggest and ugliest crevasse yet. More than twelve feet wide in spots, its mouth opens to a depth of ten feet in places, to the top of a ragged ice bridge. The bridge has a disturbing grayish cast that makes it look anything but solid.

Our chests heaving as we gulp the cold air, we halt our uphill climb. Safety team members map out a route, and we ignore fears that we've tested this glacier one too many times. I can't help thinking about Max Demorest.

For several hundred yards, we push and pull the Hotsy parallel to the glacial scar. We're headed toward a spot where the crevasse opening is narrower, about six to eight feet wide, with a bridge one foot below the glacier surface. Several team members test the bridge and declare it solid, but we all know that the true test will be the Hotsy passing over it.

We move toward the potential crossing, then point the Hotsy sled uphill on the rope team's command. Perpendicular to the crevasse, we make a full-power, full-throated charge. Driving our feet into the ice and our shoulders into the ladders and ropes, we plow toward the abyss. I grasp the forward ladder rungs in a white-knuckle hold, partly to push with all my strength and partly to be sure that I'm holding on to something if the bridge gives way.

As we begin to cross, the front metal curls of both ladder skis slam into the lip of ice at the far side of the crevasse. The ladder tips bend backward, threatening to break off, but that's not our main worry.

We've stalled atop the ice bridge.

Commands ring out from front and back: "Keep going!" "Lift the front end!" "Don't stop!" Fierce growls we once feared from polar bears now come from us. We push as one, forcing the nearly half-ton machine up and out of the bridged-over crevasse. Several members of the ladder brigade stumble as we gain speed. They hang onto the rungs and are dragged across the last few feet of the bridge. It holds.

On the far side of the crevasse we pause, allowing our spent muscles to relax. Relieved smiles creep onto our faces. The worst is

EXPEDITION TEAM MEMBERS PUSH THE HOTSY UPHILL OVER A CREVASSE. *(U.S. COAST GUARD PHOTOGRAPH BY JETTA DISCO.)*

behind us. Slowly but steadily we crest the hill; then we quicken our pace upon catching sight of the orange flags. The last leg seems almost easy. We erupt in whoops and cheers at BW-1.

Jim checks his watch: 7:56 a.m. A trek that we thought would take four hours has taken less than two, with no injuries and no damage to the Hotsy. As we hang onto the ladders or sprawl on the ice, Steve reflects on his military career. "I'm thinking about Iraq, Pakistan, doing things with a pretty elite group of guys. And *that* was amazing," he says. "From WeeGee's inventiveness to the team effort, any Special Forces unit would have been proud of that accomplishment."

Jim marvels at how two ladders that Lou bought last-minute in Keflavík turned out to be among our most critical equipment. "Where there's a will," he says, "there's a way."

THE BREAK IS brief, as moving the Hotsy is only part of the job. We need to haul more fuel, hoses, ropes, and other equipment to BW-1 for WeeGee to start melting. Several of us hike back to base camp with the ladder skis and cram the necessary supplies into a large Pelican case. When we're done, it weighs more than four hundred pounds, so we attach ropes to drag it up the glacier. We're tired from the Hotsy move, there are fewer of us working, and the surface has grown slushy since dawn. Even with the ladder skis it feels like pulling a reluctant donkey up a hill, and we anoint ourselves "the Mule Team." What follows is a two-hour torment, complete with loud and imaginative curses cast on everyone who isn't helping us.

Doc plays a leading role on the Mule Team, pulling with the strength of a much younger man and entertaining us with stories from his youth and his Coast Guard travels. Only later does he reveal that he nearly didn't make the team.

For days, safety leaders have cautioned that we've become too casual on the glacier, ignoring risks and tempting fate by failing to rope ourselves together. Doc tells us that when he left BW-1 after the Hotsy move, alone and unroped—a double mistake in the safety team's view—the glacier opened beneath his feet. Luckily, he threw out his arms and halted his fall at his armpits. His first thought, he says, was to get out fast. But he admits that his immediate motive wasn't to save himself; he didn't want to hear Nick say, "I told you so."

Doc's drop turns out to be the closest the glacier comes to claiming any of us.

A happier adventure befalls Jetta on her way to base camp after the Hotsy move. Walking with her head down, Jetta glimpses something dark on the glacier surface about a hundred yards from BW-1. She kneels and carves it from the ice with help from Jim and Rob. It's a piece of frozen fabric, striped blue and gray, about the size of her palm. It looks and feels unlike the clothing that expedition members wear, and no one has reported torn or lost gear. It's a long shot, but Jetta preserves the cloth for testing, in the hope that it might be from the Duck's fabric-covered wings. Months might pass before an answer, but her discovery raises hope of good news to come.

WEEGEE STARTS MELTING the first hole at 1:00 p.m. He chooses a spot where John and Frank drilled yesterday with the auger. He straddles the trench, keeping the black steel pipe vertical and aiming the boiling water from the Hotsy downward into the ice. WeeGee plans to burrow nearly twice the radar-reported depth of the anomaly, to be sure not to miss anything. The rest of us stand around watching, wishing the hole would open faster.

When the pipe is several feet into the ice, WeeGee looks up and notices that Jaana's hands are bare.

"You have gloves?" he asks.

"Yes," she says, quizzically.

"Put 'em on."

When she does, WeeGee steps aside and hands her the hot pipe. It's an act of appreciation, a tribute to her discovery. The work is slower than expected, the glacier resists the intrusion, and after a while WeeGee reclaims his place above the hole. At 2:15, he reaches a depth of sixty feet. Hand over hand, he pulls up fifty feet of black hose attached to the ten-foot pipe and sets it aside.

Alberto goes to work with his camera, a 4-mm lens surrounded by tiny high-intensity lights encased in a silver shell the size of a ripe pear. The camera hangs from a thick black wire, and Alberto unspools enough to reach the bottom of the hole. He drops it in, turns on the video screen, crouches on the ice, and drapes a coat over his head to block the glare. He pulls up the camera a foot at a time, searching the screen for any hint of the Duck. After several minutes, Alberto stands and runs his hands through his curly black hair. He stares at the hole and says nothing. We tamp down

ROBERT "WEEGEE" SMITH MELTS THE FIRST HOLE AT BW-1. WATCHING ARE (FROM LEFT) JIM BLOW, ALBERTO BEHAR, KEN HARMAN, AND MITCHELL ZUCKOFF. *(U.S. COAST GUARD PHOTOGRAPH BY JETTA DISCO.)*

our disappointment, knowing that if a hole is even a foot from the Duck the camera might see nothing. And because the Duck might be nose-down or nearly vertical in the ice, it would make a narrow target.

This might take many holes and lots of time, everyone agrees, so nine team members return to base camp to eat lunch and begin packing. Eight of us remain at BW-1: WeeGee, Lou, Jim, Jaana, Doc, Frank, John, and me. WeeGee starts a second hole, this one at the exact spot where Jaana placed the flag over the anomaly. But after forty minutes, with the hole about forty-five feet deep, the hose starts to spray water from where it's attached to the Hotsy. WeeGee turns off the Hotsy to repair what he diagnoses as a blown rubber O-ring. Doc shuts off a portable Honda generator that powers a pump that draws water into the Hotsy. The glacier reclaims its quiet majesty.

I notice that Jim has wandered some fifty yards away. He stands with his back to us, facing Koge Bay, talking on a satellite phone. After several minutes, he turns and walks toward us, his shoulders slumped. We gather around as he says, "Everyone's got to get off the ice." Confused, we look at each other and back at Jim. He explains that rough weather is approaching faster than expected, so Air Greenland is sending its two biggest helicopters to airlift all of us and as much gear as possible before nightfall. The first helicopter will be here in an hour.

We thought we'd have a full day or more at BW-1, and we've brought lights to work through the night. We anticipated driving fifteen, twenty, or more holes to probe the anomaly that every one of us can picture in our minds from Jaana's radar screen. Now we have an empty first hole, an unfinished second hole, and an order to leave.

It feels like a cruel joke. Everything for naught: Lou's relentlessness and sacrifices; Jim's labors and dedication; Jaana's persistence; WeeGee's inventiveness; our hard work and shared joy from the unexpected results at BW-1; the time and money spent; the triumph of the Hotsy move. This was the Duck Hunt's best shot, and now it's apparently over.

As Jim's words sink in, Lou's face goes slack. Normally he doesn't hesitate to question the Coast Guard commander's orders, but Jim looks as dejected as anyone. Lou and the rest of us say nothing.

Before we disperse to gather our personal gear, WeeGee breaks the silence: "I'm staying. Grab my stuff and have the last helicopter pick me up here."

To my surprise, Jim doesn't object. I see an opening, so I look to WeeGee. He gives me a slight nod.

"I'll stay to help," I say.

Jim hesitates, then approves. Lou raises his eyebrows and shoots me a look with one possible meaning: Are you sure you know what you're doing?

Not really, but I'm staying anyway.

Five minutes later, WeeGee and I are alone on the ice at BW-1, the final two searchers atop the Koge Bay glacier for the final hours of Duck Hunt 2012.

WASTING NO TIME, WeeGee replaces the O-ring in the hose and tries to restart the generator. It refuses to turn over, so he takes it apart, checking the oil, the connections, the spark plug, the filters, the carburetor, the fuel cap, everything he can think of. Few people know as much about engines, and the generator is a pretty basic machine. Yet no matter what he tries, WeeGee can't restart it. Without a way to draw water into the Hotsy, we can't melt holes. Our effort feels cursed.

WeeGee calls base camp on the walkie-talkie, asking that a backup generator be dropped off to us when the first helicopter arrives. Frustrated, I start breaking down equipment to shove into the Pelican case. WeeGee carries Alberto's camera and the case containing the video screen to the unfinished second hole. It's about fifteen feet short of the desired depth, but with nothing better to do, WeeGee wants a look. It's 4:00 p.m.

He covers his head with a black puffy coat to view the screen and feeds the camera to the hole's bottom. I watch him as I work, hoping. WeeGee pulls up the camera as patiently as a fisherman testing his line. After a few feet he stops. I hold my breath.

"Hey, Mitch," he calls, "come here. Take a look at this."

I race to his side, drop to my knees, and duck under the puffy coat. Our heads almost touching, his right shoulder against my left, WeeGee points to the bottom right corner of the screen. It's unmistakable, a sight so beautiful, so satisfying, so perfect, yet seemingly so impossible that I blink several times to be sure: a black plug with a wire extending from it, with a white band wrapped around the wire.

My eyes dart around the screen. I spot a cable on the opposite side of the hole. Nearby are objects that look like fuses. Rivets. More wires. We see dark shadows just beyond the camera's view that promise more vintage World War II aircraft parts where they don't belong: under thirty-eight feet of ice, on a glacier several miles from Koge Bay, in almost the precise spot where a 1943 military report says a rescue plane called a Grumman Duck, serial number V1640, crashed on November 29, 1942, with three heroes aboard.

As WeeGee and I stare at the screen, we see the final pieces of the puzzle that reveals what happened that fateful morning seventy years ago. Under the original plan, John Pritchard and Ben Bottoms would have landed the Duck, then hiked to the PN9E to get Bill O'Hara and Paul Spina. By the time they returned with the injured men, visibility likely would have been too poor to take off safely, and they would have waited for the weather to clear. But Max Demorest's fall into the crevasse changed everything. When Lolly Howarth ran to the Duck with the terrible news, Pritchard and Bottoms knew that waiting wasn't an option. They hustled Howarth aboard the Duck and took off immediately for the *Northland*, to collect ropes and tools and able-bodied men for an emergency rescue attempt. Pritchard bravely flew into the teeth of the storm, lost his bearings, called for guidance, then slammed into the glacier at the exact spot where WeeGee and I kneel.

We came within a hairbreadth of failure. With the Hotsy idle and time running out, if we had bored the hole a few feet in a different direction, we might have been standing atop the Duck without ever knowing it.

WeeGee and I throw our arms around each other, both of us grinning like new fathers.

"We've got it," WeeGee says.

Two hours later, darkness is falling and the storm is bearing down. There's no time for the helicopter pilot to shut down the engines, so WeeGee and I rush aboard under the spinning blades. We're met by cheers, hugs, and backslaps. Our walkie-talkie call informing base camp—WeeGee made sure that Lou was the first to hear—had the predictable effect on the team: disbelief, followed by jubilation.

Now, everyone on this last flight to Kulusuk crowds around to see the camera images I made from the video screen. Smiles spread from one to the next as they witness what we came for: hard evidence of the Duck's crash site. The more we analyze the images and the circumstances, the more certain we are. Everything adds up: the depth of the discovery, which matches the predicted ice accumulation of seventy years; the precise coordinates of the 1943 crash report; the absence of any other known plane crash on this area of the glacier; metal and electrical parts found in a Grumman Duck. Add that to a radar hit showing a large under-ice anomaly and signals from the magnetometer. Proof positive.

We'll have to return to Koge Bay, ideally next summer, with heavy equipment to excavate perhaps fifty tons of ice atop the plane to reach the bodies of John Pritchard, Benjamin Bottoms, and Loren Howarth. But we've solved the mystery of where they've been all these years. They had to go out, and now they can come back.

Aboard the helicopter, Lou fights tears: "I'm just so happy for Nancy."

Jim can't stop smiling. "Everything that went wrong," he says, "it's like it was supposed to happen. It's like divine intervention." Our handshake expands into a bear hug.

Watching from his seat, WeeGee shoots me a grin.

"Hey, Mitch, how does it end?"

"Like this, WeeGee. Like this."

═══ EPILOGUE ═══

AFTER GREENLAND

1943–PRESENT

THE FIRST HALF of 1943 was a busy time for war news, from the German surrender at the Battle of Stalingrad, to U.S. troops' capture of Guadalcanal, to the Warsaw Ghetto uprising. Yet when the press blackout lifted, the crashes and rescues in Greenland became a momentary sensation.

The biggest splash occurred in May 1943, after the U.S. Army issued a lengthy press release describing the extraordinary events of the previous six months. Newspapers across the country, including *The New York Times*, ran page-one stories based on the military's account. The *Los Angeles Times* enhanced its coverage with an exclusive interview with native son Armand Monteverde. Uncomfortable talking about the experience, Monteverde said his goal was to resume ferrying bombers, "preferably in the South Pacific."

Coinciding with the press release, Monteverde, Harry Spencer, and Don Tetley went to the White House on May 3, 1943, where they met with President Franklin D. Roosevelt. They emerged from the meeting spit-shined and smiling for an official photograph with General H. H. "Hap" Arnold.

Several days later, newspaper readers nationwide awoke to a twelve-part syndicated series written by Oliver La Farge, an Army Air Forces captain who'd won the Pulitzer Prize for fiction in 1930. The series focused on the PN9E crash and aftermath, mentioning the crashes of John Pritchard's Duck and Homer McDowell's C-53 almost in passing. Later, La Farge's series became part of a book called *War below Zero*, written with none other than Bernt Balchen and the writer Corey Ford.

The popular radio program *The Cavalcade of America* turned the PN9E saga into a twenty-five-minute radio play called *Nine Men against the Arctic*. The fictionalized account featured cheesy studio sound effects; the crunch of feet on snow sounded suspiciously like a man squeezing corn starch in a leather pouch. Worse yet was the stilted dialogue, which put the turgid in dramaturgy. Consider this imagined exchange between Monteverde and Spencer in the cockpit shortly before the crash:

Monteverde (From California)
   You know, Spence, I don't like this place.
Spencer (Texas Drawl. Aged 22)
   It sure is a long way from Texas.
Monteverde
   You can't see anything. All this whiteness everywhere.
   No horizon. How high are we, anyhow?
Spencer
   Reckon we're plenty high, but you can't be sure. It's like
   flying through milk.

Not long after the radio play, the story of the Greenland crashes faded from view. Like most of the men and women who served during World War II, the survivors and rescuers rejoiced at the war's end and returned to ordinary lives. In doing so, they joined a generation that endured terrible threats and remarkable events, only to tuck away their memories with their old uniforms.

ALTHOUGH HE BORE the brunt of blame for the PN9E crash in official reports, Armand Monteverde received the Legion of Merit for his actions during the months that followed. His citation credited him with "high devotion to duty and complete disregard for his own safety" in caring for his crew after the wreck. Legion of Merit medals also were given to the six other PN9E survivors and Don Tetley.

ARMAND MONTEVERDE IS WELCOMED HOME BY HIS SISTER ADA LEHR (LEFT) AND HIS MOTHER, VIRGINIA MONTEVERDE, WITH NIECE DEANNE LEHR ON HIS SHOULDERS. *(U.S. ARMY PHOTOGRAPH.)*

While on leave in California after being rescued, Monteverde enjoyed a brief moment of celebrity. Newspaper photos showed him being greeted by his beaming mother and sister, with his seven-year-old niece on his shoulders, wearing his new captain's hat.

After recuperating, Monteverde returned, as he had hoped, to ferrying planes for the Air Transport Command. He continued his service during the Korean War, and spent twenty-two years in the air force before retiring as a lieutenant colonel. Along the way he married and had a son. Armand Monteverde died in California in 1988. He was seventy-two.

Like Monteverde, Harry Spencer also was promoted to captain during his time on the ice. Afterward, he too continued to ferry bombers for the Air Transport Command. In August 1943, Spencer wrote a letter to leaders of the Boy Scouts of Dallas explaining all that he'd been through in Greenland. In it, he credited God and his Eagle Scout training for his survival. The letter made one request: "I have not been where I could pay my dues," Spencer wrote. "If you can tell me what that amount is, it would be a favor to me, as I would like to be connected with Scouting always."

After the war, Spencer opened a hardware store in Texas with his brother-in-law, then launched a successful air-conditioning business. Everything he might have imagined about his life when he fell into the crevasse came true. He and his wife, Patsy, had two daughters, Peggy and Carol Sue; a son, Tommy, who died in childhood of leukemia; and three grandchildren. When Patsy was diagnosed with multiple sclerosis, Spencer dedicated himself to caring for her.

He served as a city councilman in Irving, Texas, as director of a local hospital, and as district commissioner of the Boy Scouts of America. He was a board member of the local branch of the Girl Scouts and served on the boards of the Irving Chamber of Commerce, the Texas Commerce Bank, and the Irving YMCA.

Spencer taught Sunday school at his Methodist church for thirty-five years. He won the Distinguished Irving Civic Award and the High-Spirited Citizen Award for Extraordinary Contributions to the City of Irving, and was named Rotarian of the Year, among other honors.

Spencer's family knew him as warm and funny, and they'd remember him as a man who bought toilet paper in bulk long before warehouse stores. When his younger daughter Carol Sue asked why, Spencer explained: "I have been without toilet paper," he told her, "and I am never going to be without toilet paper again!"

At Carol Sue's urging, in 1989 Spencer returned to Greenland to visit the site of the PN9E crash. He wrote afterward that he was motivated by a desire to revisit "the pristine whiteness of the Ice Cap snow, which seems to have no dimension, the crystal blue of the bay water, and the lurking shadows of the crevasses [which] hold me in deep awe of God's wonderful creation."

After flying over the crevasse-laden field where his bomber went down and where he nearly died, Spencer asked the helicopter pilot to land at the site of the long-gone Motorsled Camp. "As I stood in the sunny Arctic silence and looked south toward the bay, it all came back to me," he wrote. "The view that was etched in my mind all the years came into view. I can't tell you the feelings that came to me. Nothing can equal that moment."

Spencer stepped onto the ice cap carrying two items: an American flag and a small plaque with the names of Max Demorest, Clarence Wedel, John Pritchard, Benjamin Bottoms, and Loren Howarth. It read, "In Memory: Five valiant men. They gave their lives in effort to save others."

"When we took off," Spencer wrote later, "the American flag and plaque remained on that vast sea of silence and nothingness to mark the spot of our experience so long ago."

Harry Spencer died in Texas in 2004. He was eighty-three. Carol Sue Spencer Podraza wanted to fill her father's tombstone with his war record and his many other honors and achievements. But Spencer had extracted a promise from her that his stone would bear only the insignia of the Boy Scouts, because its oath had guided him through boyhood to Greenland and beyond. Harry Spencer's tombstone reads, "On my honor, I will do my best, to do my duty to God and my country. To help other people at all times, to obey the Scout Law, and to keep myself physically strong, mentally awake, and morally straight."

LOSING BOTH FEET seemed to make William "Bill" O'Hara even tougher. After recovering from surgery, frostbite, and other

HARRY SPENCER ON THE ICE CAP WITH A FLAG AND PLAQUE HONORING THE LOST MEN.
(COURTESY OF CAROL SUE SPENCER PODRAZA.)

injuries, O'Hara graduated with honors from Georgetown University Law School. After a brief time in a wheelchair, he was fitted with prosthetic limbs, and for as long as he lived O'Hara never returned to the chair. Even after a late-in-life stroke, he resented needing a cane.

Although he was a lifelong New Deal Democrat, O'Hara harbored lasting bitterness about not being invited to the Roosevelt White House along with Monteverde, Spencer, and Tetley. He blamed the president. "Dad was still in a wheelchair then," his daughter Patricia said, "and Roosevelt refused to have someone in a wheelchair there. Dad said, 'That son of a bitch is in a wheelchair himself!'"

At the end of the *Cavalcade of America* radio play, O'Hara spoke briefly on the air as a special guest. "With the passage of time," he said, "I have regained some of the one hundred pounds I lost in weight, and the experience grows more and more unreal, a bad dream that one wants to forget. The only reality now is the reality of the day-by-day winning of the war."

For two years after his return from Greenland, O'Hara refused to see his girlfriend Joan, feeling as though he was no longer the man she'd fallen in love with. But she persisted. They married, and together they had a son, three daughters, and eleven grandchildren. O'Hara became a clerk in the Lackawanna County, Pennsylvania, Commissioners' Office, and spent three terms as the county's register of wills. He also served a dozen years on Pennsylvania's Public Utility Commission before opening a law practice in Scranton.

Years after Greenland, O'Hara tried to find his fellow PN9E crew members but was stymied by military rules. He complained to a reporter: "The Army has some screwy regulation that it won't divulge the addresses of veterans." In time, though, he succeeded in reaching several. He and some of the other PN9E crewmen crossed paths now and again, but for the most part they had little contact after their ordeal.

On several occasions, journalists sought out O'Hara to discuss his wartime experiences. He'd oblige, to a point: "All I have left is the pain and suffering," he told a reporter in 1982. "I can recall it being a son-of-a-bitch for eighty-eight days." That was as deeply as he'd reflect for public consumption.

"I haven't dwelled on what happened to me and the others forty years ago," O'Hara said. "I've been too busy for that kind of stuff."

His eldest daughter remembered him as an Irish charmer, a straight-talking man who liked several drinks better than one, a father who could be difficult but also supremely capable, a "tough nut" who wouldn't let Greenland get the best of him. William "Bill" O'Hara died in Pennsylvania in 1990, at seventy-two.

PAUL SPINA RETURNED home to upstate New York and worked odd jobs before becoming a factory foreman at the Chicago Pneumatic Tool Company. He loved to fish and to drink beer, ideally at the same time. He never lost his good cheer. "Anytime it was a bad situation," said his daughter, Jean, "he would turn it around and make you feel better about it."

Spina told her about his Greenland adventure and later shared the story with her children. Occasionally he'd pull out a yellowed scrapbook filled with newspaper clippings and mementos to relive it himself. Spina didn't need the scrapbook to remind him, though. He remained susceptible to the cold, and at times his hands and arms swelled and ached. Spina dismissed it as arthritis, but his wife and daughter believed that a more accurate diagnosis was Greenland.

Paul Spina died in New York of a sudden heart attack in 1978, at age sixty-one. His family blamed that on the ice cap, too.

Alfred "Clint" Best recovered fully from his physical and mental distress. After the war, he graduated magna cum laude with an accounting degree from Baylor University. Best went to work for Dow Chemical and rose through the ranks during a thirty-five-year career before retiring in 1970. He was active in his church and president of the Tulsa Rose Society. Never an athlete himself, Best coached Little League baseball because his son, Robert, was on the team.

After retiring, Best became involved in community service, above all helping underprivileged children at a local elementary school. "They called him 'Kinderpa,'" Robert Best said. "He loved children, loved getting on the floor, making arts and crafts."

In addition to Robert, Best and his wife, Amy, had two daughters, eleven grandchildren, and two great-grandchildren. Alfred "Clint" Best died in Texas in 2002, at eighty-four.

VASTLY DIFFERENT FUTURES awaited the two PN9E crewmen plucked from the ice cap on November 28, 1942, during the Duck's one successful rescue flight.

Alexander "Al" Tucciarone returned home to the Bronx and went to work in the metal-polishing business. He married his fiancée, Angelina, to whom he'd written the postcard shortly before the crash that said, "Will see you soon." They had a son and a daughter, two grandchildren, and a full life together.

Every winter, the cold settled painfully in Tucciarone's hands, reminding him of the crash and the weeks that followed. For almost three decades, Tucciarone refused to fly. He relented in 1971, to attend a ceremony for the dedication of two buildings at the U.S. Coast Guard Aviation Training Center in Mobile, Alabama. One building was living quarters for bachelor officers, named Pritchard Hall. The other was the enlisted men's barracks, named Bottoms Hall.

"For those two," Tucciarone said at the time, "I would do anything."

Before the ceremony, he met the mothers of the Duck's pilot and radioman. Tucciarone found himself at a loss for words: "How can I tell them what's in my heart, how I feel about being alive, while their sons who saved my life are dead?"

Alexander "Al" Tucciarone died in New York in 1992, at age seventy-seven. Twenty years later, his son choked up talking about him. "He was a great guy—everything about him," Peter Tucciarone said. "His personality, his attitude, everything. His appreciation of life and people, the whole nine yards. . . . I'm so grateful for the two men who saved his life. He lived for another fifty years after that, and if not for them, I wouldn't be here talking to you."

TUCCIARONE AND LLOYD "Woody" Puryear stayed in touch as they moved through different hospitals and were assigned to different bases. In a letter to Tucciarone in April 1943, Puryear wrote that he was awaiting orders for a new posting, but in the meantime "the circulation still isn't back to normal in my feet yet, and I'm still very sensitive to cold. The gangrene in my toes hadn't quite reached the bone, so they were able to save them."

THE MOTHERS OF JOHN PRITCHARD AND BENJAMIN BOTTOMS UNVEIL PLAQUES WITH THEIR SONS' NAMES AT THE DEDICATION OF COAST GUARD RESIDENCE HALLS IN THEIR HONOR. *(U.S. COAST GUARD PHOTOGRAPH.)*

Puryear congratulated Tucciarone on being promoted to sergeant and signed off, "A Greenlander Buddy."

Eight months later, Puryear's sister wrote to Tucciarone's wife with solemn news. Puryear was en route from a base in Montreal to Walter Reed Army Medical Center, "ill with a lung ailment caused I'm sure from the exposure last year. . . . The [doctor] talked so discouraging to me, and poor little thing was getting ready to come to Kentucky on furlough for Xmas, and [was] so thrilled. This is about to kill us."

Puryear died within weeks, in January 1944. His obituary called him "Campbellsville's first World War II hero" and said, "Taylor County buried one of its favorite and most beloved native sons Tuesday afternoon in one of the largest and most impressive military funerals ever held from a local church." Lloyd "Woody" Puryear was twenty-six.

FOR HIS RESCUE flights in the PBY Catalina, Bernard "Barney" Dunlop earned a Distinguished Flying Cross, as did his copilot, Nathan Waters. Dunlop's medal citation captured the drama of the mission, honoring him for "brilliant" flying "under rigorous Arctic conditions, at the risk of his own life, the lives of his crew, and disabling damage to his plane."

After the war, Dunlop served as a lawyer for the Nassau County government, on Long Island, New York. He and his wife had one child, their daughter Nancy. He died in New York in 1964, at fifty-four.

A Distinguished Flying Cross also went to B-17 pilot Kenneth "Pappy" Turner for keeping the PN9E crew alive and fed. It went unnoticed, but Turner probably deserved another medal for talking sense into Monteverde, Spina, and Best when they considered hurling themselves into a crevasse. Turner was promoted to major in July 1944. After the war, he returned to his former life as an airline pilot. He died in California in 1994, at ninety-one.

AFTER THE RESCUES, Don Tetley was happy to leave behind Greenland and the life of an enlisted man. He went to officer candidate school in Miami and was commissioned a second lieutenant in the Army Air Forces. After the war, he returned to San Antonio and launched a career as an electrician. Don Tetley died in Texas in 1986, at sixty-nine.

FOR HIS WORK organizing and overseeing the rescues, Bernt Balchen received a Distinguished Flying Cross. Despite the heroic contributions by Dunlop in the PBY Dumbo and Turner in the B-17, among others, Balchen received the lion's share of acclaim. The *Chicago Tribune* declared in a headline: "Bernt Balchen Saves 7 on Ice Cap." *The New York Times'* editorial page rhapsodized about him in an item headlined "Flier of the Snows." It read, in part, "Once more Bernt Balchen . . . has been the hero of a rescue by air in the kind of country which his physical equipment, experience, and mechanical skill have made his own, the Greenland Ice Cap. . . . The story of the fight of the crew against injuries, frostbite and the foulest of weather reads like a chapter out of romance. To bring them aid and finally save the survivors, Balchen had to use every trick of his long years in the polar wastes."

Balchen barely had time to catch his breath before he saw more action. Upon his return to Bluie East Two, Balchen received secret orders to wipe out a German weather station on the northeast coast of Greenland. Turner's B-17 was still at the base, so the bomber finally got an opportunity to fulfill its destiny. A combat crew joined Turner's usual team, and Turner piloted the mission with Balchen standing in the cockpit, supervising. Even though Balchen thought the station was deserted, they followed orders and strafed it with incendiary rounds, setting it ablaze.

Balchen left Greenland before the war ended. His next role was commander of an air transport system credited with evacuating more than three thousand Norwegians, Americans, and others from Sweden. He also aided the Norwegian underground by transporting tons of supplies and communications equipment from Scotland and England to occupied Norway. After the war, Balchen cofounded the Scandinavian Airlines System and commanded the Alaska-based 10th Rescue Squadron of the U.S. Air Force, among other exploits.

Later, Balchen returned to the limelight by fueling doubts about Admiral Richard Byrd's claim to have flown over the North Pole in 1926. Based on his calculations, Balchen concluded that Byrd had turned back well short of the pole. He was alternately reviled and hailed for the assertion. Disagreements still fester over Byrd's North Pole claim and Balchen's role in efforts to debunk it.

Bernt Balchen died in New York in 1973, a week before his seventy-fourth birthday. He was buried with full military honors at Arlington National Cemetery, strangely enough in a grave next to Byrd's. In 1999, on the centennial anniversary of Balchen's birth, the U.S. Congress passed a resolution honoring him for "a lifetime of remarkable achievements . . . [and] extraordinary service to the United States."

MAX DEMOREST'S DEATH in a crevasse was marked by obituaries in publications as diverse as the journal *Science*, his hometown newspaper in Flint, Michigan, and *The New York Times*. The tribute in *Science* hailed the young professor turned lieutenant for discoveries about the motion of glaciers that provided "the key to the solution of a baffling problem" that bedeviled scientists for more than a century.

As often happens in science, some of Demorest's conclusions were later disputed. But no one challenged the obituary writer's closing lines: "In the death of Max Demorest both glaciology and glacial geology have lost a master mind who, even before the age of thirty-two, brought clarity where there had been much confusion. He will be remembered by his colleagues as one who did not engage in disputation, who by his calm, convincing reasoning caused no rancor nor ever lost a friend."

Three more tributes followed: Demorest's wife, Rebecca, who'd been his research assistant, became a geologist in her own right, working for the U.S. Geological Survey in Washington; his daughter, Marsha-Jo, followed him to the University of Michigan, where she studied botany and geology; and finally, a glacier in Greenland was named in Demorest's honor.

In 1947 War Department officials wrote to the six remaining PN9E survivors, asking for details that might be used to retrieve Demorest's body. Harry Spencer's reply put an end to that idea. He wrote, "Crevasses which we observed had no bottom either to sight or falling objects. . . . Even if the remains of the airplane were found to establish a local area, the particular crevasse into which Demorest fell would, by this time, have shifted, closed or frozen over, making the discovery of his body impossible."

FOR SEVERAL YEARS after Clarence Wedel's death, his father feared that his eldest son's rejection of the family's religious beliefs had denied him salvation. Jacob Wedel prayed on it, with no relief. Then he had a dream about a rosebush in his garden that never bloomed. In the dream, he was about to dig up the bush when a divine voice told him to leave it. When the bush bloomed, the voice said, it would prove that Clarence was in heaven. Jacob Wedel woke comforted. In time the bush did bloom, producing the most beautiful roses in his garden.

Clarence Wedel's wife, Helen, never remarried. Their daughter, Reba, went to Washington State University, became an English teacher, then left teaching to become a full-time wife and mother of two daughters. Later in life she became a watercolorist, winning local and national awards and showing in galleries.

Clarence Wedel's body was never found, but his parents put his name on a headstone they knew would eventually bear theirs. Under Wedel's date of birth, the inscription carved in granite reads, "Passed from Earth to Glory, December 7, 1942, While on a Rescue Mission in Greenland, Where He Now Rests, Awaiting a Glorious Resurrection."

LOREN HOWARTH'S LOCAL newspaper wrote a stirring obituary when news of his death was revealed. "War in all its shattering bitterness was impressed upon Wausaukee this week with the announcement of the death of Corporal Loren Howarth," the item began. The obituary listed the schools he'd attended, his military record, and the details of the crash that were known at the time. It concluded: "The sadness and sacrifice of war has fallen on a mother who now must carry Wausaukee's first gold star. . . . Taps for Corporal Loren Howarth, a fine boy and a hero."

In 1951, a VFW post in Illinois was named for Howarth. His wife, Irene, who'd been his landlady while he was at college, remarried her ex-husband after Loren's death. Family members say his parents grieved his death for as long as they lived.

THE COAST GUARD ship from which John Pritchard and Benjamin Bottoms alighted was assigned a new Duck and flying crew. The *Northland* spent the rest of the war on the Greenland Patrol, its endless routine occasionally interrupted by momentary drama. A highlight came in September 1944, when the ship chased a Nazi vessel through ice floes for seventy miles off Great Koldeway Island. With nowhere to run, the German crew scuttled their ship, and the *Northland* took eight officers and twenty-eight enlisted men prisoner. The *Northland* ended the war with two Battle Stars.

After the war, the ship was supposed to be sold for scrap, but was instead purchased by American Zionists. Renamed the *Jewish State*, the ship was enlisted in a brave but unsuccessful attempt to break the British blockade preventing the transport of Jews from Europe to Palestine. Upon the birth of Israel in 1948, the *Jewish State* was renamed the *Eilat* and became the first warship in the new Israeli navy. Later it became a barracks ship.

At a decommissioning ceremony in 1962, the chief of Israeli naval operations gave the ship a fond sendoff: "Our old sister, the scrap heap, has served faithfully. She was a symbol of our fight, without arms or speed. We part from her with mixed feelings." Parts of the original *Northland* are on display in a museum in Israel.

The U.S. Coast Guard honored the ship in 1984 by bestowing the name *Northland* on a new 270-foot cutter that today patrols the Atlantic, the Caribbean, and the Gulf of Mexico.

THEIR FAMILIES NEVER forgot John Pritchard Jr. or Benjamin Bottoms. Nor were they forgotten by the men they served with, or by those who followed them in service.

The most visible memorials to their heroism are the residence halls named in their honor at the Coast Guard facility in Alabama. Their names also are inscribed in granite on the U.S. Coast Guard Aviation Memorial in Elizabeth City, North Carolina. Atop the memorial is a quote from the book of Exodus: "I bore you on eagles' wings and brought you to myself."

Countless other, less public, tributes have been made to their heroism and their sacrifices. And as long as there is a U.S. Coast Guard, there will be countless more. John Pritchard and Ben Bottoms are honored every time Coast Guard rescuers go out, knowing that they might not come back.

BACK AT THE Hotel Kulusuk, on the last day we were all together, I pulled a dollar bill from my wallet and asked for the signatures of my sixteen fellow expedition members. It's my own Short Snorter membership certificate, and it's one of my most prized possessions.

When Lou saw me passing around the bill for signatures, he wanted to join the club. He smiled, shrugged, and turned his palms skyward: "Can I borrow a dollar?"

Three days after we discovered the crash site, after a dozen members of the expedition had gone home and the storm had passed, I returned to the glacier with four others: WeeGee, Jim, Lou, and Jetta. We melted more holes at BW-1, bringing our total to twelve, and located more wires and other objects from the Duck, all at thirty-eight feet under the surface. We had hoped to pinpoint the position and orientation of the fuselage, and perhaps locate the bodies, but frozen hoses and issues with the Hotsy intervened.

Later, Jim determined that the first pieces WeeGee and I saw were "the wing rib from the reinforced section of either the upper or lower wing, near the fuselage, where cables attach to stabilize and reinforce the wings." What we originally thought was a fuse box was most likely the Duck's "Large Junction Box," designed to allow four electrical cables to pass through the wing. Best of all, the objects were located perhaps two feet from the plane's fuselage. Jaana's flag placement above the anomaly had been perfect.

At this writing, Jim and Lou are planning a follow-up expedition to recover and repatriate the remains of Pritchard, Bottoms, and Howarth. Jim's superiors at the Coast Guard are on board, working closely with Joint POW/MIA Accounting Command. We've been e-mailing and talking regularly, and when the follow-up expedition happens, I intend to be there. Next time, whether we need it or not, we'll have a Snublebluss. The polar bear warning system that Lou ordered from England was waiting for him upon his return from Greenland. Meanwhile, WeeGee has taken to calling our against-the-odds discovery "the miracle on ice."

Lou is still sorting through the bills from the 2012 expedition and making plans for 2013 and beyond. After completing the Duck Hunt, Lou has his eye on a mission memorialized on another one of his patches: locating and recovering the C-53 with the bodies of McDowell and his crew. To Lou's credit and to the benefit of my credit rating, he sent me a chunk of the money from the Coast Guard contract to help cover my American Express bill.

Nancy Pritchard Morgan Krause was thrilled about our findings. But the actuarial tables for a woman her age tempered her excitement: "I hope I'm still here to see John brought home." I hope so, too.

THE 2012 DUCK HUNT EXPEDITION TEAM: (BACK ROW, FROM LEFT) RYAN SAPIENZA, ROB TUCKER, MITCHELL ZUCKOFF, ALBERTO BEHAR, MICHELLE BRINSKO, FRANK MARLEY, NICK BRATTON; (MIDDLE ROW, FROM LEFT) JETTA DISCO, LOU SAPIENZA, JIM BLOW, TERRI LISMAN, STEVE KATZ, JAANA GUSTAFSSON; (FRONT ROW, FROM LEFT) W. R. "BIL" THUMA, KENNETH "DOC" HARMAN, ROBERT "WEEGEE" SMITH, JOHN BRADLEY. *(U.S. COAST GUARD PHOTOGRAPH BY JETTA DISCO.)*

CAST OF CHARACTERS

(In Alphabetical Order)

HISTORICAL

BERNT BALCHEN—A Norway-born aviation pioneer, Balchen became a colonel in the U.S. Army Air Corps during World War II and commanded the northernmost American base in Greenland. He was the first to spot the downed B-17 PN9E and was instrumental in planning and executing rescue efforts.

ALFRED "CLINT" BEST—A tech sergeant and cryptographer at Bluie West One, Best volunteered to take part in the search for Homer McDowell's missing C-53 aboard the B-17 PN9E.

BENJAMIN BOTTOMS—A Coast Guard radioman from Georgia, Bottoms volunteered to serve alongside pilot John Pritchard Jr. during the Duck's fateful mission to rescue the B-17 PN9E crew.

MAX DEMOREST—A gifted glaciologist with a doctorate from Princeton University, Demorest sidetracked his academic career during the war to become commanding officer of a rescue outpost on the east coast of Greenland called Beach Head Station.

HENDRIK "DUTCH" DOLLEMAN—Born in the Netherlands, Dolleman was a sergeant in the U.S. Army who spent time in Antarctica before being sent to Greenland to take part in rescue efforts.

CHARLES DORIAN—An ensign aboard the *Northland* and later a Coast Guard captain, Dorian alerted the ship's captain to the sighting of the lost Canadians. He served as a primary source and technical adviser on this book.

BERNARD "BARNEY" DUNLOP—A U.S. Navy lieutenant, Dunlop was the pilot of the PBY Catalina/Dumbo used in the rescue efforts at the Motorsled Camp. His crew included co-pilot Nathan Waters, flight engineer Alex Sabo, radioman Harold Larsen, and Dr. P. W. Sweetzer, medical officer at Bluie East Two.

WILLIAM EVERETT—A corporal who was a crew member on the C-53 cargo plane that crashed in Greenland on November 5, 1942.

RICHARD FULLER—A Coast Guard ensign aboard the *Northland* who volunteered to go ashore to lead what he thought would be a two-week rescue effort. He and his fellow Coast Guardsmen were stuck on the ice for five months.

DAVID GOODLET—Pilot of the Royal Canadian Air Force A-20 that crashed in Greenland on November 10, 1942.

JOSEPH HEALEY—A sergeant, Healey served on the crew of Admiral Byrd's 1933–1935 Antarctic expedition before becoming a dogsled driver on rescue teams in Greenland during World War II.

FRANK HENDERSON—A navy ensign who was pilot of the backup PBY Catalina during the rescue mission.

LOREN "LOLLY" HOWARTH—Radioman on the B-17 PN9E that crashed on November 9, 1942, while searching for Homer McDowell's lost C-53. Howarth's work rebuilding the broken radio was credited with making possible the B-17 crew's survival. He was aboard the Duck on November 29, 1942.

THURMAN JOHANNESSEN—A private who was a crew member on the C-53 cargo plane that crashed in Greenland on November 5, 1942.

HERBERT KURZ—Navigator on Kenneth "Pappy" Turner's supply B-17.

EUGENE MANAHAN—A staff sergeant who was a crew member on the C-53 cargo plane that crashed in Greenland on November 5, 1942.

HOMER MCDOWELL—A captain who was pilot of the C-53 cargo plane. His crash on the east coast of Greenland on November 5, 1942, set in motion the search-and-rescue efforts by the B-17 PN9E and the Grumman Duck.

J. G. MOE—An army captain who served as navigator on Jimmie Wade's unsuccessful attempt to rescue the survivors at the Motorsled Camp.

ARMAND MONTEVERDE—A captain from California, Monteverde was pilot of the B-17 PN9E that crashed on November 9, 1942, searching for Homer McDowell's lost C-53. Monteverde's leadership was credited with helping his crew survive after the crash.

AL NASH—Navigator of the Royal Canadian Air Force A-20 that crashed in Greenland on November 10, 1942.

WILLIAM "BILL" O'HARA—A stoic native of a coal-mining town in Pennsylvania, O'Hara was navigator of the B-17 PN9E that crashed on November 9, 1942, while searching for Homer McDowell's lost C-53.

FRANCIS "FRANK" POLLARD—A lieutenant commander, Pollard was the captain of the U.S. Coast Guard cutter *Northland* during the rescue of the Canadian crew and the efforts to rescue the PN9E crew.

JOHN PRITCHARD JR.—South Dakota–born and California-raised, the eldest of five children, Pritchard became a Coast Guard flier after more than two years as an enlisted man in the navy. As a lieutenant assigned to the cutter *Northland*, Pritchard piloted the ship's amphibious rescue and surveillance plane, the Grumman Duck. After leading the successful rescue of three Canadian fliers, Pritchard attempted to rescue crew members of the downed B-17 PN9E.

LLOYD "WOODY" PURYEAR—A staff sergeant from Kentucky, Puryear was a volunteer searcher aboard the B-17 PN9E that crashed on November 9, 1942, while searching for Homer McDowell's lost C-53.

EDWARD "ICEBERG" SMITH—A rear admiral with a PhD in oceanography from Harvard, Smith led the Coast Guard's Greenland Patrol during the war and played a role in planning rescue efforts for the PN9E crew.

HARRY SPENCER—An effervescent Texan and a natural leader, Spencer was copilot of the B-17 PN9E that crashed on November 9, 1942, while searching for Homer McDowell's lost C-53. Along with Armand Monteverde, Spencer was credited with keeping alive fellow crew members after the crash.

PAUL SPINA—An easygoing private from upstate New York, Spina was the engineer on the B-17 PN9E that crashed on November 9, 1942, while searching for Homer McDowell's lost C-53.

WILLIAM SPRINGER—A lieutenant, he was copilot of the C-53 cargo plane whose crash in Greenland on November 5, 1942, set in motion the search-and-rescue efforts by the B-17 PN9E and the Grumman Duck.

HAROLD STRONG—A captain who'd made a fortune in the stock market and worked as a trapper in Alaska, Strong led a search-and-rescue team in Greenland that played an important role in the effort to rescue the downed B-17 PN9E crew.

DON TETLEY—A sergeant from Texas with a background as a cowboy, Tetley served on the motorsled rescue team with Max Demorest. He spent more than two months with survivors of the PN9E crash.

ALEXANDER "AL" TUCCIARONE—A laborer and truck driver in the Bronx before the war, Tucciarone was a private who served as assistant engineer aboard the B-17 PN9E that crashed on November 9, 1942, while searching for Homer McDowell's lost C-53.

KENNETH "PAPPY" TURNER—A native of Salt Lake City, Turner was the pilot of a B-17 that braved months of terrible flying conditions to airdrop supplies to the survivors of the B-17 PN9E crash.

JIMMIE WADE—A Canadian bush pilot, Wade volunteered for an unsuccessful rescue mission to the Motorsled Camp.

ARTHUR WEAVER—Radioman of the Royal Canadian Air Force A-20 that crashed in Greenland on November 10, 1942.

CLARENCE WEDEL—A mechanic from Kansas, Wedel was a private en route to England who hitched a ride aboard the B-17 PN9E days before it crashed in Greenland.

ROBERT WIMSATT—A colonel, Wimsatt was commander of the U.S. Army's Greenland bases and took part in overseeing the rescue efforts.

MODERN

ALBERTO BEHAR—A member of the Duck Hunt expedition of 2012, Behar holds a PhD in electrical engineering and a degree in robotics. He designed the down-hole camera that played a pivotal role in the expedition. Behar also oversees one of the experiments aboard the NASA Mars rover *Curiosity*.

AARON BENNET—An independent television producer, Bennet served as Lou Sapienza's de facto business partner, handling financial, media, and other duties while he pitched a show based on the exploits of Lou and his exploration company.

JAMES "JIM" BLOW—A U.S. Coast Guard commander, Blow serves in the Office of Aviation Forces and was the service's liaison and mission leader for the Duck Hunt. He led the Coast Guard team during the 2012 search-and-recovery mission.

JOHN BRADLEY—A member of the Duck Hunt expedition of 2012 and a veteran ice rescue guide, Bradley served as a leader of the safety team. He oversees the mountaineering department of REI's flagship store in Denver.

NICHOLAS "NICK" BRATTON—A member of the Duck Hunt expedition of 2012 and a former mountaineering guide, Bratton served as a member of the safety team. Away from the ice, Bratton designs land conservation programs in Washington State. Bratton also serves as the board treasurer for the Fallen American Veterans Foundation.

MICHELLE BRINSKO—A member of the Duck Hunt expedition of 2012, Brinsko served as the base camp manager and chief cook. If there had been a morale officer, she would have been it. Away from Greenland, she is a physical therapist.

JOE DEER—A Coast Guard commander who served in the Office of Aviation Forces, Deer was a key member of the service's team that launched the Duck Hunt.

JETTA DISCO—A member of the Duck Hunt expedition of 2012, Disco is a petty officer second class in the Coast Guard. She serves in the service's public affairs office in New York. Her images of the expedition constitute the official record.

JAANA GUSTAFSSON—A member of the Duck Hunt expedition of 2012, Gustafsson was the team's inexhaustible ground-penetrating radar expert. A Finn living in Sweden, Gustafsson has a PhD in geophysics and works as a land surveyor.

KENNETH "DOC" HARMAN—A member of the Duck Hunt expedition of 2012, Harman is a captain in the Coast Guard and a flight surgeon. Harman has responded for the Coast Guard to trouble spots around the globe.

JERRY HOWARTH—Nephew and primary next of kin of Loren "Lolly" Howarth.

STEVE KATZ—A member of the Duck Hunt expedition of 2012, Katz served as second-in-command of North South Polar. A retired colonel in the Army Reserves, Katz works as an executive in a logistics and transportation company.

TOM KING—A retired Coast Guard captain, King was active in the Coast Guard Aviation Association when he suggested that efforts be made to locate the lost Grumman Duck and recover the bodies of the men aboard the plane.

NANCY PRITCHARD MORGAN KRAUSE—Sister and primary next of kin to John Pritchard Jr. She was an inspiration to and a stalwart supporter of the Duck Hunt.

TERRI LISMAN—A member of the Duck Hunt expedition of 2012, Lisman is an image scientist for the National Geospatial-Intelligence Agency. Brought on the mission by the Coast Guard, she operated the magnetometer used to confirm BW-1 as a key search location.

JOHN LONG—Retired master chief petty officer in the U.S. Coast Guard and an indefatigable researcher into the crash of the Grumman Duck.

FRANK MARLEY—A member of the Duck Hunt expedition of 2012, Marley served as leader of the safety team. A third-year medical student, Marley intends to practice expedition medicine. At the time of the Duck Hunt, he had just returned from Afghanistan, where he was a captain in the U.S. Army National Guard.

JAMES MCDONOUGH—A lieutenant colonel with the U.S. Department of Defense Prisoner of War/Missing Personnel Office, or DPMO, who was a primary contact for Lou Sapienza as he planned the Duck Hunt.

EDWARD "BUD" RICHARDSON—Stepson of Benjamin Bottoms and an important supporter of the effort to find the Duck and the men it carried.

LOU SAPIENZA—The son of a U.S. Navy veteran from World War II, Lou spent years as a commercial photographer before volunteering for three missions to Greenland to find a lost P-38 fighter plane known as *Glacier Girl*. Eventually that led him to a years-long effort to find the lost Grumman Duck. He led the 2010 and 2012 recovery expeditions to Greenland.

RYAN SAPIENZA—A member of the Duck Hunt expedition of 2012, Ryan Sapienza served as keeper of the expedition log and aide-de-camp. He is Lou's son.

ROBERT "WEEGEE" SMITH—A member of the Duck Hunt expedition of 2012, Smith oversaw logistics and operated the Hotsy water pressure unit used to explore anomalies beneath the glacier at Koge Bay. A veteran of Greenland expeditions in search of World War II planes, he builds and repairs rally race cars in Vermont.

MARC STORCH—Cousin by marriage of Loren "Lolly" Howarth and keeper of Howarth's Legion of Merit.

DONALD TAUB—A retired Coast Guard captain who served in Greenland, Taub played a major role in researching the events from November 1942 to May 1943, and served as a historical adviser on this book.

W. R. "BIL" THUMA—A member of the Duck Hunt expedition of 2012, Thuma served as an authority on geophysics. Thuma is a consultant who markets technology for natural resource exploration around the world.

ROB TUCKER—A member of the Duck Hunt expedition of 2012, Tucker is a lieutenant commander in the Coast Guard, based in Washington. A pilot, Tucker served as second-in-command to Jim Blow on the Coast Guard expedition team.

ACKNOWLEDGMENTS

IT'S A POWERFUL and unsettling experience to be drawn into the orbit of someone possessed by an impossible dream. At times I wondered if Lou Sapienza would awake and abandon his quixotic plan to find three airmen entombed in a glacier. Or maybe he'd suffer one too many sacrifices and surrender to self-preservation. But no matter how many setbacks Lou faced, nothing deterred him. The Duck Hunt expedition was the accomplishment of a rare and remarkable man. If I'm ever lost, I hope that Lou decides that I need to be found.

I'm profoundly indebted to Commander Jim Blow of the U.S. Coast Guard. He made countless essential contributions to the Duck Hunt and to this book, and he welcomed me on the mission with respect and kindness. I admire his leadership and value his friendship. As I say elsewhere, he's a gentleman.

Deep thanks to all my expedition mates, several of whom offered valuable comments on the manuscript. I hope that the friendships we made will endure as long as the memories. In alphabetical order: Alberto Behar, John Bradley, Nick Bratton, Michelle Brinsko, Jetta Disco, Jaana Gustafsson, Ken Harman, Steve Katz, Terri Lisman, Frank Marley, Ryan Sapienza, Bil Thuma, and Rob Tucker. A special shout-out to my friend Robert "WeeGee"

Smith, with whom I shared one of the most extraordinary moments of my life.

Retired master chief petty officer John Long was a tireless researcher and an unflagging advocate for the families of John Pritchard, Ben Bottoms, and Loren Howarth. This book is marbled with his insights and contributions.

Three retired Coast Guard captains played key roles in this book. Don Taub spent years investigating these events: he tracked down participants, analyzed innumerable documents, and corrected mistakes made in earlier accounts. Thomas C. King Jr., who kick-started the Duck Hunt, provided essential help as I began this project. Charles Dorian sent me the rare photographs he took as an ensign aboard the *Northland* during the fall of 1942. His tales of life aboard ship were invaluable.

I'm thankful to the family members of the heroes whose stories are told here. Trusting me with their loved ones' reputations, they shared documents, photos, and insight into these remarkable men. Thank you, Nancy Pritchard Morgan Krause, who brought to life her late brother, John Pritchard; Edward "Bud" Richardson, who reached into his childhood memories to describe his stepfather, Benjamin Bottoms; and Marc Storch and Jerry Howarth, who shared stories of Loren Howarth.

Thanks also to Pete Tucciarone, who told me loving anecdotes about his father, Alexander "Al" Tucciarone; Reba Greathead and Eric Langhorst, who enlightened me about her father and his grandfather-in-law, Clarence Wedel; Robert Best, who made me feel as though I knew his father, Alfred "Clint" Best; Jean Gaffney, daughter of Paul Spina, who shared his priceless forty-page account of his ordeal; Patricia O'Hara, daughter of William "Bill" O'Hara, who regaled me with tales of his toughness; Nancy Dunlop, daughter of PBY pilot Bernard Dunlop; and Carol Sue Spencer Podraza, daughter of Harry Spencer, whose vivid stories and delightful way of telling them made every conversation a pleasure.

Among the historians, librarians, and archivists who helped me were two who went above and beyond: Coast Guard historian Robert M. Browning Jr., PhD, and William H. Thiesen, PhD, Atlantic Area Coast Guard historian. Thanks also to Martina Soden of the Scranton, Pennsylvania, Public Library; Karen Kortbein of the Wausaukee, Wisconsin, Public Library; and Mark C. Mollan, archivist in the Navy/Maritime Reference section at the National Archives and Records Administration. My graduate assistant, Sarah Testa, devised a filing system for my research that imposed order on chaos. Thanks to Evan Caughey for creating a sparkling website.

Chuck Greenhill, owner of the last flying Grumman Duck model J2F-4, and Duck enthusiast Bill Floten shared their deep knowledge of the wonderful little plane. In Aaron Bennet, Lou Sapienza found a true partner and a match for his relentlessness. Deep thanks to Jim and Nancy Bildner, who generously supported this effort.

Every author should have Richard Abate as an agent. Actually, scratch that, because then I'd have to wait longer for his wise counsel and great humor. His contributions to this book and to my career are too many to list. There's no female equivalent of *mensch*, so I propose *claire*, as long as it's first applied to my editor, Claire Wachtel. I'm only sorry that she didn't join me on the ice.

In my last book I thanked Jonathan Burnham at HarperCollins for everything; now double that. Double it again for Michael Morrison. If she'd abandon her allegiance to the Giants, publicist extraordinaire Kate Blum would be perfect. Kathy Schneider, Tina Andreadis, and Leah Wasielewski move heaven and earth with grace and charm. Miranda Ottewell's unerring eye kept my copy clean. Doug Jones and his team are the best in the business. Special thanks to Melissa Kahn at 3Arts and to Elizabeth Perrella at HarperCollins, for taking such good care of me and my work.

At Boston University, I'm grateful to my students for challenging and invigorating me, and to Dean Tom Fiedler; Chairman Bill McKeen; and journalism professors Bob Zelnick, Lou Ureneck, Chris Daly, Susan Walker, Nick Mills, Elizabeth Mehren, Rob Manoff, Peter Southwick, Jon Klarfeld, Michelle Johnson, and R. D. Sahl, among many others.

I'm fortunate to have close friends who'd be first-rate company in the tail section of a B-17: confidant and coconspirator Brian McGrory; Dan Field; Colleen Granahan; Isabelle Granahan-Field; Eliza Granahan-Field, to whom this book is secretly dedicated; Bill, Ruth, and Emily Weinstein; my oldest friend, Jeff Feigelson, whose advice on the manuscript and legal matters is deeply appreciated; my partner in crime and class, Dick Lehr; Chris Callahan; Naftali Bendavid; Kathryn Altman; Helene Atwan; Joann Muller; and the late Wilbur Doctor. A nod to the memory of a loyal and loving pal named Briggs. He would have loved Koge.

The competition of our youth has mellowed, but I'll always be trying to impress my brother, Allan Zuckoff, and to meet the standards he sets. Thanks also to the extended Zuckoff and Kreiter clans. Next summer in Bethany.

My parents, Sid and Gerry Zuckoff, didn't like my traipsing around Greenland; they liked it better when they could limit my wanderings to an imaginary line near Ira Meyers's house. Yet in this project as in all things, their love and support have been the secret weapons of my life.

My daughters are my northern lights: they fill me with delight as they take away my breath. Isabel Zuckoff is a person I admire as much as I love. Eve Zuckoff is a person I love as much as I admire.

My wife, Suzanne Kreiter, rescues me daily. She makes everything possible.

NOTES

THIS IS A work of nonfiction. No liberties were taken with facts, dialogue, characters, or chronology. All quoted material comes from interviews, direct observation, reports, diaries, letters, flight logs, military documents, news stories, books, or some other source cited below. Descriptions of people and places are based on observations by the author, interviews, written materials, photographs, and newsreel images. Unless noted, the author conducted all interviews, either in person or by phone.

PROLOGUE: THE DUCK

1 "Situation grave": "Incoming Message, November 26," Papers of Corey Ford at Dartmouth College, Rauner Special Collections Library. Punctuation added.

4 eager to get going: Charles Dorian, interview, September 4, 2011.

4 not one but two round-trips: "Will evacuate remaining Seven Baker 17 (B-17) personnel via *Northland* plane in two flights," radio message, November 28, 1942.

4 less than one mile: USCGC *Northland* log entries, November 29, 1942.

4 suspended over the *Northland*'s deck: The scene on board the *Northland* on November 29, 1942, comes from Coast Guard documents on the cutter and the Duck, and Dorian interviews, September 4 and November 11, 2011. Dorian, an ensign at the time, served as a communications officer aboard the *Northland* and witnessed the Duck's takeoff that day.

5 Pritchard and Bottoms's craft: Corydon M. Johnson, "Erection and Mainte-

nance Instructions for Model Grumman J2F-4 Airplanes," August 16, 1939,
U.S. Coast Guard historical archives.

6 stripped the space: Edward F. Clark, "In the Line of Duty," *Coast Guard
   Magazine*, March 1943, p. 13.

6 sipping hot coffee: B-17 survivor Al (Alexander) Tucciarone to CBS news
   reporter Michelle Marsh, undated letter (believed to be from the 1980s),
   Benjamin Bottoms's personnel file, U.S. Coast Guard Historical Office.

6 Pritchard set the Duck's propeller: Dorian, interview, September 4, 2011.
   The description of the preflight procedures is based on Grumman's "Pilot's
   Handbook for Model J2F-4 Airplane," August 21, 1939, and a December
   2, 2011, interview with pilot Chuck Greenhill. Greenhill is an authority on
   flying a Duck and, at the time of the interview, owner of the only known
   Grumman Duck J2F-4 still airworthy.

7 almost zero: Greenhill, interview, December 2, 2011.

8 "You have to go out": Clark, "In the Line of Duty," p. 14.

1: GREENLAND

9 twenty feet or more: "The Greenland ice sheet holds enough water to raise
   the global sea level with ~7m." Rune G. Graversen et al., "Greenland's Con-
   tribution to Global Sea-Level Rise by the End of the 21st Century," *Climate
   Dynamics* 37, no. 7–8 (October 2010): 1427–42.

9 Erik the Red: Jonathan Grove, "The Place of Greenland in Medieval Icelandic Saga Narrative," in "Norse Greenland: Selected Papers of the Hvalsey
   Conference 2008," special issue, *Journal of the North Atlantic* 2 (2009): 30.

9 "good name": Ibid.

10 called his discovery Vinland: Douglas R. McManis, "The Traditions of Vinland," *Annals of the Association of American Geographers* 59, no. 4 (December
   1969): 797–814.

10 four thousand years: Eske Brun, "Greenland," *Arctic* 19, no. 1 (March
   1966): 62.

10 "a greenish tinge": Adam of Bremen, "[Greenland in] Chapter 37," in
   Beskrivelse af øerne i Nordern [Description of the Islands in the North] (Co-
   penhagen: Wormianum, 1978). Original Latin text, *Descriptio insularum
   Aquilonis*, c. 1075, and Danish translation, with commentaries by Allan A.
   Lund; English translation by B. Wallace. Also see J. Grove, "The Place of
   Greenland in Medieval Icelandic Saga Narrative," in "Norse Greenland:
   Selected Papers from the Hvalsey Conference, 2008," special issue, *Journal of
   the North Atlantic* 2 (2009): 30–51.

13 breaking both arms: John A. Tilley, "The Coast Guard and the Greenland
   Patrol," accessed January 3, 2012, www.sondy-logen.dk/images/pdf/green
   land_patrol.pdf, p. 6.

13 animal skins, seal oil, and fish: Ibid., p. 2.

13 kept Greenland isolated: U.S. Coast Guard, Public Information Division,

The Coast Guard at War: Greenland Patrol (Washington, D.C.: U.S. Coast Guard, 1945), pp. 14, 36.

14   weather in Europe: Ibid., p. 2.

14   "a war for weather": Bernt Balchen, Corey Ford, and Oliver La Farge, *War below Zero* (New York: Houghton Mifflin, 1944), p. 4.

14   milky white mineral called cryolite: Ibid., pp. 6–8.

15   "a crippling blow": U.S. Coast Guard, *Greenland Patrol*, p. 8.

16   secret preparations: Gerard Kenney, *Dangerous Passage: Issues in the Arctic* (Toronto: Natural Heritage, 2006), p. 129.

16   "enslavement, miscegenation": Ibid., p. 15.

16   icebergs in shipping lanes: Ibid., p. 4. Also see Malcolm F. Willoughby, *U.S. Coast Guard in World War II* (Annapolis, Md.: U.S. Naval Institute Press, 1957), pp. 95–110.

17   "Phooey on Bluie": The complete poem is included in the World War II scrapbook of Paul J. Spina, provided by his daughter, Jean Spina Gaffney.

17   in July 1942: The story of the Lost Squadron is drawn primarily from David Hayes, *The Lost Squadron* (Edison, N.J.: Chartwell, 2008), pp. 18–54.

18   "Send women": Norman D. Vaughan, *My Life of Adventure* (Mechanicsburg, Pa.: Stackpole, 1995), p. 76.

2: "A MOTHER THAT DEVOURS HER CHILDREN"

19   November 5, 1942: This account of the crash of Captain McDowell's C-53 comes primarily from the official U.S. Missing Air Crew Report (MACR), 42–15569. Material also was taken from handwritten notes in the Papers of Colonel Bernt Balchen, U.S. Air Force Historical Agency, Maxwell Air Force Base.

19   "fifty-two miles up a fjord": Dan Ford, "Remembering Bluie West One," www.warbirdforum.com/bluie1.htm (accessed February 1, 2012).

20   the weather report: MACR 42–15569.

21   "forced landing": Balchen's notes say the crash occurred "due to one of [its] engines out of operation." But the MACR gives no reason and does not indicate whether a cause was discussed during the radio transmissions from the C-53.

21   seventy-seven total hours: "Information Requested from Form 5 Unit," form detailing pilot hours for Captain Homer C. McDowell Jr., Collection of Harry Trice, U.S. Coast Guard Historian's Office. This hour total is as of the end of October 1942.

21   ten miles away: Because the exact location of the C-53 remains unknown at this writing, this is only an estimate based on its reported position, the apparent location from which flares were fired, and the areas searched.

22   "Down on Ice Cap": "Description of Accident," MACR 42–15569. A small disagreement exists between the MACR and Balchen's notes on the initial reported altitude, with the MACR claiming 9,400 feet and Balchen writing

9,200. Either way, the C-53 gave a much lower altitude in subsequent messages.

22 B-25 Mitchell bombers: MACR 42–15569, p. 1. Also, Donald M. Taub, *The Greenland Ice Cap Rescue of B-17 "PN9E," November 5, 1942, to May 8, 1943*, monograph published by the U.S. Coast Guard History Program, www .uscg.mil/history/articles/GreenlandPatrolIceCapRescueTaub2011.pdf (accessed December 15, 2011), p. 1.

23 "I asked C-53": "Report of Search for Lost Plane (C-53)," Collection of Harry Trice, U.S. Coast Guard Historian's Office.

24 considered equally brilliant and brave by his friends: Numerous accounts exist of Demorest's great intellect and his bravery. Overall, the most complete is found in William S. Carlson, *Greenland Lies North* (New York: Macmillan, 1940), an account of a journey by Carlson and Demorest through Greenland to study air currents.

24 a doctorate from Princeton University: "Max H. Demorest, Glacier Authority," obituary, *New York Times*, December 11, 1942.

25 "I hoped that Max's ignorance": Carlson, *Greenland Lies North*, p. 21.

25 five to ten miles: Taub, *Greenland Ice Cap Rescue*, p. 2.

26 three or four days: Ibid.

26 "transmitting MOs": Aircraft Accident Classification Committee (AACC) report, C-53, #5569, document, December 19, 1942, p. 1.

27 cargo bay was empty: MACR 42–15569, p. 1–3. The report makes clear that the plane was returning to Greenland without mail, passengers, or cargo.

27 two days' rations: AACC report #5569, p. 2.

27 a deadly low of minus 10: Ibid.

27 "No luck": Onas P. Matz, *History of the 2nd Ferrying Group* (Seattle: Modet, 1993), p. 168.

27 "there was little probability": AACC report #5569, p. 2.

28 the aurora borealis: William S. Carlson, *Lifelines through the Arctic* (New York: Duell, Sloan & Pearce, 1962), p. 90.

28 hypothermia: Numerous sources consulted, including MedlinePlus, www.nlm .nih.gov/medlineplus/hypothermia.html; Rick Curtis, "Outdoor Action Guide to Cold Weather Injuries," www.princeton.edu/~oa/safety/hypocold.shtml; and Mayo Clinic health information sheets, www.mayoclinic.com/health/hypother mia/DS00333/DSECTION=symptoms (all accessed January 28, 2012).

3: FLYING IN MILK

30 the same day: Harry Spencer, handwritten air log, November 1942.

31 a B-17F: Some reports describe the PN9E as a B-17E, but the official crash report lists it as a B-17F. For details on the B-17F, see National Museum of the U.S. Air Force factsheet, www.nationalmuseum.af.mil/factsheets/fact sheet.asp?id=2453 (accessed January 11, 2012). See also Bill Yenne, *B-17 at War* (St. Paul, Minn.: Zenith, 2006).

32 a special oath: U.S. Centennial of Flight Commission, "The Norden Bomb-
    sight," www.centennialofflight.gov/essay/Dictionary/NORDEN_BOMB
    SIGHT/DI145.htm (accessed January 27, 2012).

32 *Cyanide for Hitler*: These names were all used on B-17s by the 452nd Bomb
    Group, based in England during World War II. See www.angelfire.com
    /ne2/b17sunriseserenade/452ndnames.html (accessed January 11, 2012).

33 ferrying crew: MACR 42–5088, p. 1.

33 their first foreign mission: Paul J. Spina, unpublished memoir, found by his
    daughter, Jean Spina Gaffney, p. 3.

33 "Goodbye, sea food": Spina, memoir, p. 3.

35 Pikiutdlek: Wallace Hansen, *Greenland's Icy Fury* (College Station: Texas
    A&M University Press, 1994), p. 44.

35 at low altitudes: Harry E. Spencer Jr., "Report on Crash of B-17 No. 5088
    and Subsequent Operations," typewritten version provided by Donald M.
    Taub. A version of this account was published in Matz, *History of the 2nd
    Ferrying Group*, pp. 139–52.

35 number-four engine lost oil pressure: Spina, memoir, p. 5.

35 two men walked over: Ibid., p. 6.

36 a bet among themselves: Ibid.

36 "nice, warm sack": Ibid.

37 went into the cockpit: Ibid., p. 7

37 about seven thousand feet: Oliver La Farge, *The Long Wait*, published as a
    series of newspaper articles by the Army Air Forces Aid Society, distributed
    in 1943 by King Features Syndicate, chap. 1. Much of the material in this
    serial was reprinted in the book *War below Zero*, so the two are almost inter-
    changeable.

38 a horizontal line of blue sky: Harry Spencer, speech to Irving, Texas, Rotary
    Club, August 31, 1989. A videotape was provided to the author by Spencer's
    daughter, Carol Sue Spencer Podraza.

38 happens on the ground: La Farge, *Long Wait*, chap. 1.

4: THE DUCK HUNTER

40 Walking double-time through baggage claim: This scene was wit-
    nessed by the author, as explained in the text. From this point forward,
    events and comments witnessed firsthand will not be cited in the
    source notes.

41 "to limit the loss": DPMO website, www.dtic.mil/dpmo/about_us/ (accessed
    December 13, 2011).

44 More than eighty-three thousand: DPMO website, www.dtic.mil/dpmo
    /summary_statistics (accessed December 13, 2011).

46 only three served in the Coast Guard: William H. Thiesen, "Lieutenant
    Thomas James Eugene Crotty: A Coast Guard Leader, Hero and Prisoner
    of War," *Bulletin* (USCGA Alumni Society), June 2008, pp. 17–18. Thiesen

confirmed in an e-mail to Commander Jim Blow on October 10, 2012: "As far as our records indicate, Pritchard and Bottoms are the only other unrecovered MIAs in service history."

5: A SHALLOW TURN

48   a mild manner: Balchen, Ford, and La Farge, *War below Zero*, p. 62.

49   working nights in a gas station: Spina, memoir, p. 9.

49   seven hundred hours of flight time: U.S. Army Air Forces (USAAF) aircraft accident report, document, April 19, 1943, p. 1.

49   sensitive to the feelings of others: Balchen, Ford, and La Farge, *War below Zero*, pp. 62–63.

50   eight thousand feet: USAAF accident report, p. 2.

51   Pilots in Greenland told stories: La Farge, *Long Wait*, chap. 1.

51   trusted their guts: Spencer says he and Monteverde believed there was enough clearance to turn. Matz, *History of the 2nd Ferrying Group*, p. 140. Unless otherwise noted, details of the PN9E crash come primarily from affidavits given by Monteverde and Spencer, included in the MACR as part of the military's investigation.

51   one thousand feet of clearance: Spencer, speech.

51   no man would have objected: In their MACR affidavits, Spencer and assistant engineer Alexander Tucciarone both say that, to the best of their knowledge, everyone aboard thought they were well above the ice cap.

52   laborer and truck driver back home in the Bronx: World War II enlistment records, from www.fold3.com/page/86088102_alexander_l_tucciarone (accessed January 23, 2012). See also "Ferry Tales," *Sunday Morning Star*, January 10, 1943.

52   postcard that read: Postcard addressed to "Miss A. Imperati," November 2, 1942, provided to the author by Peter Tucciarone.

52   couldn't see five feet beyond the bomber: Tucciarone to Marsh, p. 2.

53   "Somebody pull me in—I'm freezing": Burlyn Pike, "Nineteen Days of Freezing Hell," *Courier-Journal Roto Magazine*, n.d., in Lieutenant John Pritchard, Coast Guard personnel file.

53   five foot four: Paul J. Spina, military ID, located in his personal scrapbook. His draft record at ancestry.com (accessed January 20, 2012) lists him at five foot three.

53   one of the volunteer searchers: The injuries sustained by the crew are detailed in the MACR affidavits given by Armand Monteverde, Harry Spencer, and Alexander Tucciarone. There is disagreement whether Puryear was also thrown through the PN9E's nose, but most accounts suggest he was not.

55   about 10 degrees: Spencer, MACR affidavit, p. 1.

55   flamed out: La Farge, *Long Wait*, chap. 1.

55   dry, sandy snow: Ibid.

56   "lack of depth perception": USAAF accident report, p. 1.

57   afraid to even try it: Spencer's account in Matz, *History of the 2nd Ferrying Group*, p. 141.

58   When engineer Paul Spina regained his senses: Spina, memoir, p. 7.

58   admired the small man's toughness: "Fortress Pilot Tells of 148 Days on 'White Hell' Icecap," *Los Angeles Times*, May 5, 1943.

58   marveled that the tarp was in the plane: Spina, memoir, p. 7.

59   "Memories of home": Pike, "Nineteen Days."

59   parched: Balchen, Ford, and La Farge, *War below Zero*, p. 48.

59   brought along a thermos: Alexander L. Tucciarone, 27th Ferrying Squadron, statement taken by Charles G. Conley, Major, A.G., December 25, 1942, in Benjamin Bottoms's Coast Guard personnel file, p. 2.

59   Spina's hands were too frozen: Spina, memoir, p. 8.

60   "Am I missing you all right?": "Five Months on the Ice Cap," *Bureau of Naval Personnel Information Bulletin*, October 1943, p. 47.

60   Monteverde made a modest announcement: La Farge, *Long Wait*, chap. 1. See also Tucciarone to Marsh, p. 3.

61   whenever the storm died down: Tucciarone, statement, p. 1.

61   aspiring actor: Loren Howarth's enlistment record, www.ancestry.com (accessed January 24, 2011).

62   thirty-six days: Count of K and D rations contained in Spencer's account in Matz, *History of the 2nd Ferrying Group*, p. 140.

62   stretch the rations for ten days: Spina, memoir, p. 10. There is some disagreement about when the rations were found. In his memoir, Spina says it wasn't until after the third day.

63   a few squares of chocolate: Tucciarone to Marsh, p. 3.

63   "stoved up": Spina, memoir, p. 8.

63   suffered the most: Tucciarone, statement, p. 2.

63   Spina fished out a cigarette: Spina, memoir, p. 8.

63   a bond with copilot Harry Spencer: Spina, memoir, p. 9.

65   sixteen C-47s and six B-17s: AACC report #5569, December 19, 1942, p. 2.

65   30 degrees below zero: "Captain Tells of 148 Days in Greenland's Ice," *Chicago Daily Tribune*, May 5, 1943.

65   someone would be out looking for them: Spina, memoir, p. 8.

65   knelt together to pray: Ibid., p. 10.

65   decided to have a look around: There are multiple accounts of Spencer's fall into the crevasse, some with conflicting details. The account here relies primarily on Spencer's own version, found in Spencer's MACR affidavit, p. 2; along with Matz, *History of the 2nd Ferrying Group*, p. 141; Spina, memoir, p. 10; and La Farge, *Long Wait*, chap. 2.

66   graduated from the University of Scranton: "Attorney William O'Hara Dies; Was an Ex-PUC Commissioner," obituary, *Scranton Times*, December 26, 1990. See also Patricia O'Hara, interview, August 16, 2012.

66 see the water: "Captain Tells of 148 Days."

67 paddle along the coast: Spencer, speech.

6: MAN DOWN

68 a goner: Balchen, Ford, and La Farge, *War below Zero*, p. 48.

69 his own obituary: Harry E. Spencer Jr. obituary, *Dallas Morning News*, www
    .findagrave.com/cgi-bin/fg.cgi?page=gr&GRid=79596093 (accessed Febru-
    ary 2, 2012).

69 didn't drop for five seconds: Time estimates are based on "Speed, Dis-
    tance, and Time of Fall for an Average-Sized Adult in Stable Free Fall
    Position," www.greenharbor.com/fffolder/speedtime.pdf (accessed Febru-
    ary 3, 2012).

69 about one hundred feet from the surface: Matz, *History of the 2nd Ferrying
    Group*, p. 141.

70 Spencer landed on his back: Balchen, Ford, and La Farge, *War below Zero*,
    p. 48. In his telling of Spencer's story, La Farge does not specify exactly
    how Spencer landed. However, he writes that Spencer "brushed off the
    snow" before he stood up, which logically suggests that Spencer was on
    his back.

71 strangely serene: Spencer's official affidavit did not describe his feelings and
    vivid descriptions about his predicament, but he shared them with La Farge
    for *The Long Wait*. They are included in Balchen, Ford, and La Farge, *War
    below Zero*, p. 48–49, which is the source of much of the information about
    Spencer's fall. Spencer also described the fall in his videotaped speech to the
    Irving Rotary Club on August 31, 1989.

71 "God must have a plan for me": Spencer, speech.

71 called for help: Pike, "Nineteen Days."

71 "Get rope!": Tucciarone to Marsh, p. 4.

72 a telltale sign: Tucciarone, statement, p. 2.

72 braided six lines: The rescue of Harry Spencer is told in several places,
    sometimes with varying details. The detail about six shroud lines, for in-
    stance, appears in his MACR affidavit but not in Matz, *History of the 2nd
    Ferrying Group*, or La Farge, *Long Wait*.

72 under his armpits: Spencer, affidavit, p. 2.

74 they prayed as a congregation: Spina, memoir, p. 10.

74 "I don't even know": Ibid., p. 13.

74 butt ends of the plane's machine guns: Balchen, Ford, and La Farge, *War
    below Zero*, p. 50.

75 rubbed O'Hara's feet: Ibid.

75 sprinkled sulfa powder: Spina, memoir, p. 13.

75 hours-long shifts: Tucciarone, statement, p. 1.

75 learned to leave their gloves outside: Balchen, Ford, and La Farge, *War below
    Zero*, p. 57.

75 "Don't wear tight shoes": U.S. Army Air Forces Arctic Survival Manual, located at http://arcticwebsite.com/USAAFsurvival.html (accessed February 6, 2012).

76 blisters the size of tennis balls: Spina, memoir, p. 8.

7: A LIGHT IN THE DARKNESS

79 two hours after they left Newfoundland: This account of the crash and rescue of the RAF A-20 and its crew is derived largely from C. B. Wall, "Fourteen Days of Hell on an Icecap," *Maclean's*, May 15, 1943, a narrative based on interviews with David Goodlet and Arthur Weaver. Other sources include "Greater Love Hath No Man; A Story of the U.S. Coast Guard," *St. Petersburg Times*, May 8, 1943; Taub, *Greenland Ice Cap Rescue*, pp. 2–3; and the Coast Guard's official service history of the USCG cutter *Northland*.

80 "Good show, old cock!": Wall, "Fourteen Days of Hell," p. 9.

83 seventeen miles: Coast Guard message, November 18, 1942: "Men have left plane and are walking toward Anortek (Anoretok) Fjord. Present position of men is seventeen miles northeast of plane position."

84 less than two days old: Coast Guard message, November 23, 1942, which says in part, "*Northland* aerial reconnaissance this afternoon found snowshoe tracks of 3 men leading from westward . . . believe less than 2 days old."

87 "I just saw a light": Dorian, interview, June 25, 2012.

88 "twilight between sanity and insanity": Wall, "Fourteen Days of Hell," p. 10.

89 "Dave had his wife": Ibid., p. 94.

90 "Lieutenant Pritchard's intelligent planning": Medal citation at http://militarytimes.com/citations-medals-awards/recipient.php?recipientid=30361 (accessed February 21, 2012).

90 map salvaged from the cockpit: Spina, memoir, p. 12.

91 "Well done": SOPA Smith to *Northland*, November 24, 1942, Coast Guard message.

8: THE HOLY GRAIL

94 about $22 million: Defense Prisoner of War / Missing Personnel Office (DPMO) Operation and Maintenance, Defense-Wide Fiscal Year (FY) 2012 President's Budget, http://comptroller.defense.gov/defbudget/fy2012/budget_justification/pdfs/01_Operation_and_Maintenance/O_M_VOL_1_PARTS/O_M_VOL_1_BASE_PARTS/DPMO_OP–5_FY_2012.pdf (accessed January 31, 2012), p. 405.

95 King grew up: Thomas C. King Jr., interviews, August 21, 2011, and January 24, 2012.

95 piloting a combat rescue helicopter: Steve Vogel, "Bearing Reminders of Terror, USS *Cole* Is Back in Action," *Washington Post*, December 25, 2003.

98 "see if you can find something": John Long, interview, August 15, 2011.

98   "Lou" Sapienza spent his childhood: Lou Sapienza, interviews, including a lengthy discussion of his background on January 25, 2012.
100 dogsled leader: Wolfgang Saxon, "Norman Dane Vaughan, 100; Went to Antarctica with Byrd," *New York Times*, December 27, 2005.

9: SHORT SNORTERS

104 They salvaged what they could: Spina, memoir, p. 11.
104 a primitive calendar: Ibid., pp. 31–32.
105 nearly impossible for searchers: Report of Aircraft Accident, Form #14, Covering the Aircraft B-17F, #42–5088, Incl. #6, listing date of search, number of flights, and origin of search planes.
105 Daily logs: "Communications Relative to Lost C-53, 42–15569," memo attached to Report of Aircraft Accident #42–5088, pp. 3–4.
105 better than anyone else on earth: Pike, "Nineteen Days."
106 prayed the rosary daily: "Fortress Pilot Tells of 148 Days."
107 already sent his Christmas cards: Pike, "Nineteen Days."
107 "Short Snorters": Ibid.
107 traced its origins: Numerous accounts exist of the origins and rules of the Short Snorters, with varying details. See John T. Bills, "Meet the Flying Short Snorters," *Miami Herald*, May 31, 1942, www.shortsnorter.org /Meet_The_Flying_Short_Snorters.html (accessed February 7, 2012). See also General Mark W. Clark, *Calculated Risk* (New York: Enigma, 2007), p. 28.
107 tattooed a dollar bill on his chest: Dee Breden, "The Short Snorter Racket," *New York Times Magazine*, September 9, 1945, p. 98.
108 "a billion dollar racket": Ibid.
108 President Franklin D. Roosevelt: "President Carries 'Card' in 'Short Snorter' Club," *New York Times*, February 13, 1943.
109 radio equipment: Balchen, Ford, and La Farge, *War below Zero*, pp. 50–51.
109 glass vacuum tubes: Spina, memoir, p. 11.
109 Born in a log cabin built by his logger father: Background information on Howarth comes from interviews with Jerry Howarth, nephew of Loren Howarth, and Marc Storch, a cousin by marriage, on February 18, 2012.
109 Single when he'd enlisted: Loren E. Howarth, World War II enlistment record, www.ancestry.com.
110 the plane's batteries: Spencer, MACR affidavit, p. 2.
110 gas-powered generator: Spina, memoir, p. 12.
110 Spina cringed as he heard Howarth: Ibid.
111 "I can't do it": Pike, "Nineteen Days."
111 studied torn and incomplete assembly diagrams: "Fortress Pilot Tells of 148 Days."
111 snow poured in: Pike, "Nineteen Days."
111 rivets began to pop: Balchen, Ford, and La Farge, *War below Zero*, p. 50.

112 ham radio operator: SOPA (Senior Officer Present Afloat) to B-17 PN9E, November 27, 1942, Coast Guard message.

112 filled in the blanks: Ibid.

112 there was the manual: Spina, memoir, p. 11.

112 too excited to talk: "Fortress Pilot Tells of 148 Days."

112 "We got 'em!": Pike, "Nineteen Days."

113 felt like kings: Tucciarone, statement, p. 1.

113 Monteverde captured the crew's feelings: "Fortress Pilot Tells of 148 Days."

10: FROZEN TEARS

114 November 19 through 23: Report of Aircraft Accident, Form #14, Covering the Aircraft B-17F, #42–5088, Incl. #6 listing date of search, number of flights, and origin of search planes.

114 last few biscuits: Spina, memoir, p. 12.

115 They started with one-dollar bills: Carol Sue Spencer Podraza, daughter of Harry Spencer, interview, April 4, 2012. Podraza quoted her father telling her the story with great amusement.

115 Bernt Balchen: This profile of Balchen is derived from numerous sources, including the Bernt Balchen Papers at the Library of Congress, http://lcweb2.loc.gov/service/mss/eadxmlmss/eadpdfmss/2009/ms009032.pdf; Balchen's authorized but ghostwritten autobiography, *Come North with Me* (New York: E. P. Dutton, 1958); and "Bernt Balchen, Explorer and Pilot in Arctic, Dead," *New York Times*, October 19, 1973.

117 modest, even shy: "Bernt Balchen Weds Former Schoolmate," *New York Times*, October 21, 1930.

118 a chance to demonstrate his rescue skills: Associated Press, "Balchen Assists in 2 Rescue Feats," *New York Times*, August 7, 1942.

119 "When you fight in the Arctic": Balchen, *Come North with Me*, pp. 228–29.

119 "tossed like a leaf in a cyclone": Ibid., p. 238.

119 a small red star: Ibid.

119 a crushed dragonfly: Ibid.

120 considered it a miracle: Ibid.

120 cargo they'd brought along: Balchen to Commanding Officer, Greenland Base Command, "Subject: Search C-53 and Rescue Operations PN9E," memo, April 30, 1943, pp. 1–2.

120 froze on their reddened cheeks: "Fortress Pilot Tells of 148 Days."

120 One crewman grabbed a bundle: Balchen, Ford, and La Farge, *War below Zero*, p. 53.

121 would be joining them on the ice: Tucciarone, statement, p. 2.

121 their eyelids frozen together: Spina, memoir, p. 13. In his memoir, Spina mistakenly says this happened on Thanksgiving, when in fact Balchen first dropped supplies to the PN9E crew two days before, on November 24, 1942.

121 "Take only in small quantity": Ibid.

122 offered a piece of advice: Balchen, "Subject: Search C-53," p. 2.

122 tore a page from his diary: Balchen, *Come North with Me*, p. 240.

122 as wide as fifty feet across: Tucciarone, statement, p. 2.

122 they made O'Hara delirious: Spina, memoir, p. 14.

122 "Situation grave": "Incoming Message, November 26," Papers of Corey Ford.

123 the priority became the B-17: SOPA Smith to the *Northland*, November 24, 1942, Coast Guard message.

123 a Norwegian fur trapper and survival expert who'd been stuck in Greenland: Hansen, *Greenland's Icy Fury*, p. 125.

124 "The Arctic is an unscrupulous enemy": Balchen, *Come North with Me*, p. 240.

11. "DON'T TRY IT"

126 "further delay will seriously endanger ship and personnel": *Aklak* to SOPA, November 25, 1942, Coast Guard message for B-17 PN9E.

126 "In view of lateness of [the] season": SOPA to *Northland*, November 27, 1942, Coast Guard message for B-17 PN9E.

126 "Extremely hazardous": *Northland* to SOPA, November 27, 1942, Coast Guard message for B-17 PN9E.

126 "Do not take risks": SOPA to *Northland*, November 28, 1942, Coast Guard message. One version of this message says, "Do not take undue risks this late in season," but the word "undue" is crossed out. Although this message is dated November 28, the response to the suggestion is dated November 27, which raises the possibility that this message's date, which is notated by hand, is incorrect.

126 "Do not, repeat not, deem it advisable": *Northland* to SOPA, November 27, 1942, Coast Guard message.

127 "I shall sell life dearly": The Creed of the U.S. Coast Guard, www.uscg.mil /History/faqs/creed.asp (accessed February 21, 2012).

127 "frozen feet, a touch of gangrene, high fever": "Incoming Message, November 26."

129 Pritchard's younger brother Gil was a B-17 pilot: "Ex-'Times' Boy, Now Flying Ace, Here on Leave," *Los Angeles Times*, June 12, 1944.

129 "the touching appeal": Clark, "In the Line of Duty," p. 13.

130 a responsible, dependable boy: Nancy Pritchard Morgan Krause, interview, August 24, 2011.

131 four-tenths of a point below the bar: U.S. Rep. W. E. Evans to Rear Admiral Henry G. Hamlet, telegram, July 10, 1934, Pritchard's Coast Guard personnel file.

131 a blood test: L. C. Covell, Acting Commandant, U.S. Coast Guard, to U.S. Senator Hiram Johnson, July 30, 1934. The letter references Pritchard's positive result on the "Wassermann and Kahn blood tests," which screened for syphilis.

131 Virginia Pritchard bared her political soul: Virginia Pritchard to Administrator Chester C. Davis, July 8, 1934.

131 an accepted cadet dropped out: Rear Admiral H. G. Hamlet, Coast Guard commandant, to U.S. Senator Hiram Johnson, August 16, 1934.

131 "At reveille," Sargent recalled, "he would practically jump out of his bunk": Vice Admiral Thomas R. Sargent III (USCG Retired), commentary, *Bulletin of the U.S. Coast Guard Academy Alumni Association*, April 3, 2008.

133 "Nancy for Tick and Tick for Nancy": Krause, interview.

133 Bottoms enlisted in the Coast Guard: Benjamin A. Bottoms, Coast Guard personnel service record.

133 "Georgia Cracker": Lloyd Puryear to the parents of Benjamin Bottoms, February 19, 1943, Bottoms personnel file.

134 forced down in fog twelve miles off the Massachusetts coast: "C.G. Rescues Four Men and Plane at Sea," *Boston Globe*, December 4, 1939.

134 returned to Massachusetts with measles: Lieutenant Commander E. E. Fahey, commanding officer of Air Station Salem, Massachusetts, note, attached to Clark, "In the Line of Duty," p. 12.

134 recommended for promotion: W. N. Derby, Commandant, First Naval District, memo, October 7, 1942, Bottoms personnel file.

135 Less than twenty minutes after the *Northland*'s anchor splashed into the bay: *Northland* log, November 28, 1942; Clark, "In the Line of Duty," p. 13.

135 the Duck was in flight: By some accounts, the Duck scouted the crash site, returned to the *Northland* to confirm the plan, then took off again. This version is most credibly contained in Clark, "In the Line of Duty," in which he quotes Pollard extensively. However, most accounts have Pritchard and Bottoms going directly to the landing site, and the statements of Monteverde, Spencer, and Tucciarone do not mention an initial scouting flight.

135 flying over the crash site: Spencer, MACR affidavit, p. 3; Balchen, *Come North with Me*, p. 240.

135 canned chicken, sausages, soups, and candy: Balchen, Ford, and La Farge, *War below Zero*, p. 55.

136 took cover inside the tail section: Spina, memoir, p. 15.

136 a note that Pritchard had written while aboard the *Northland*: Ibid. See also Balchen, Ford, and La Farge, *War below Zero*, p. 55; Pike, "Nineteen Days." There are some small discrepancies among the accounts about exactly what the note said, but Spina provides the most detailed description of these events.

136 "If there's a 60-40 chance": Spina, memoir, p. 15.

137 wiped away tears: Ibid.

137 climbed atop the tail: Ibid. In Pike's version, the do-not-land signal was atop the left wing. The larger point remains, however, that the crew signaled Pritchard to fly off.

137 "Don't try it": Balchen, Ford, and La Farge, *War below Zero*, p. 55.

137 "Coming in anyway": Ibid.

137 "He won't make it, poor fellow": Pike, "Nineteen Days."

138 Several times the tail lifted: Clark, "In the Line of Duty," p. 13.

138 "You shouldn't have landed": Balchen, Ford, and La Farge, *War below Zero*, p. 56.

138 last to leave: Ibid.

139 Jesus Christ in the Greenland sky: Pete Tucciarone, interview, February 25, 2012.

139 fragile and stiff from the cold: Pike, "Nineteen Days."

139 should take their places: Spina, memoir, p. 16.

139 face-first into the snow from exhaustion: Tucciarone, statement, p. 2.

139 spaghetti waiting for him: Mignon Kilday, "Survivor Recalls Rescue," *Mobile* (Ala.) *Press*, n.d., in Bottoms personnel file.

140 clasped hands and prayed: Ibid.

140 Tucciarone heard Pritchard and Bottoms scream for joy: Tucciarone to Marsh, p. 7.

140 The ship's crew lined the rail: Clark, "In the Line of Duty," p. 13.

141 "If weather permits": *Northland* to SOPA, November 28, 1942, Coast Guard message.

142 ensure the destruction of the PN9E's Norden Bombsight: SOPA to *Northland*, November 29, 1942, Coast Guard message.

142 hearty slaps on the back: Dorian, interviews, September 4 and November 11, 2011.

142 one request: Pritchard's autograph: Clark, "In the Line of Duty," p. 13.

12: "MOs—QUICK!"

143 a flare of their own: Spina, memoir, p. 16.

143 across an active glacier: Lydia McIntosh, "Sergeant in S.A. Recalls Being Snowbound for 68 Days," *San Antonio Light*, pt. 7, May 9, 1943.

144 just such an occasion: Balchen, Ford, and La Farge, *War below Zero*, p. 57. See also Spina, memoir, p. 16.

144 "frozen feet and body poison": *Northland* to SOPA Greenland, relaying message from Tetley, December 1, 1942, message.

144 constipated: Ibid.

144 cover their sleds while they slept: Spina, memoir, p. 16.

145 taste spaghetti before Tucciarone did: Ibid., p. 17.

145 Spina knew that no one was asleep: Ibid.

145 "bright gleam of victory": Winston Churchill, *Never Give In!: The Best of Winston Churchill's Speeches* (New York: Hyperion, 2003), p. 341.

146 By noon, visibility would be less than one mile: USCGC *Northland* log entries, November 29, 1942.

146 a beautiful day on the glacier: Spina, memoir, p. 17.

146 emerged from the B-17's tail to retrieve the sleds: Spencer, statement, p. 3.

147 banquet aboard the *Northland*: Spina, memoir, p. 17.

147 "kiss the Ice Cap goodbye": Ibid.

147 the bridge gave way: This account of Demorest's fall is taken from numerous sources, including Balchen, Ford, and La Farge, *War below Zero*, pp. 59–61; Spencer, statement, p. 3; Monteverde, statement, p. 2; and McIntosh, "Snowbound for 68 Days."

148 gave him eight rolls of film: Spina, memoir, p. 18.

149 "apparently attempting to contact motor sledges in her vicinity or [at] Ice Cap Station": SOPA to *Northland*, November 29, 1942, Coast Guard message.

149 Pritchard took off in the same direction: This was long a point of contention, but in interviews with retired Coast Guard captain Donald Taub, Harry Spencer expressed certainty that Pritchard had followed the same flight path as on the previous day. Spina's account, previously unknown outside his family, also says that Pritchard flew over the wreck and headed out to sea.

149 Pritchard waggled the Duck's wings: Balchen, Ford, and La Farge, *War below Zero*, p. 60.

149 the fog grew so thick that they had to abandon their vigil: Spencer, statement, p. 3.

150 called the *Northland* for a weather report: Clark, "In the Line of Duty," p. 14.

150 "MOs, MOs—quick!": Ibid.

151 *Northland* crew members told themselves: Dorian, interviews, September 4 and November 11, 2011.

151 "Demorest and one motor sledge in crevasse": *Northland* to SOPA Greenland, November 30, 1942, message.

151 expecting that they'd be next to leave: Spina, memoir, p. 19.

151 "if by remaining, ship and personnel are endangered": SOPA Greenland to *Northland*, November 30, 1942, message.

152 "Grumman [Duck] located": *Northland* to SOPA Greenland, December 7, 1942, relaying message from Turner's B-17. See also "Activity of Airplane B17F, 41–24583, and crew in search for C-53 and search and supply of B17F PN9E," n.d., Corey Ford Papers; Taub, *Greenland Ice Cap Rescue*, p. 5.

153 one overly optimistic message from the *Northland*: *Northland* to SOPA Greenland, December 3, 1942, message.

154 "Concentrated search was discontinued": Handwritten summary attached to MACR 42–15569, pp. 2–3.

155 recommended for the Medal of Honor: "Board of Investigation—Circumstances attending loss of J2F airplane . . . recommendation for posthumous award of Medal of Honor to Coast Guard Personnel," memo, June 2, 1943, Pritchard's Coast Guard personnel file.

155 The medal was presented to his parents: "Parents Given Medal Won by Missing Flyer," *Los Angeles Times*, April 12, 1943.

155 "California Mother of the Year": "Mrs. J. A. Pritchard Named 'California Mother' for 1944," *Burbank Review*, April 24, 1944.

156 "I want to stress that I owe my life": Tucciarone, statement, p. 2.

156 "I am one of the boys whose life was saved": Lloyd Puryear to Benjamin Bottoms's parents, February 19, 1943, Bottoms's Coast Guard personnel file.

156 "I breathed a little prayer": Pike, "Nineteen Days."

13: TAPS

162 "fully resourced program": DPMO Operation and Maintenance Budget, pp. 405–6.

162 "a game changer": Lieutenant Colonel James McDonough to Lou Sapienza, e-mail, January 26, 2012.

14: GLACIER WORMS

168 Time and hardship had revealed Monteverde: Spina, memoir, p. 8.

168 considered "Lieutenant Monty" to be a hero: Ibid.

169 the third of seven children: Jean Spina Gaffney, daughter of Paul Spina, interviews, March 12 and 18, 2012.

170 he was raised a Dunkard: "McPherson Man Plunged to Death in Greenland Crevasse," *Hutchinson* (Kans.) *News-Herald*, May 27, 1943, scrapbook clipping; Eric Langhorst, Wedel's grandson-in-law, interview, March 6, 2012.

170 wrong to use his religion to avoid the war: Reba Greathead, daughter of Clarence Wedel, interview, March 13, 2012.

170 Clint Best was easygoing and introverted: Robert C. Best, son of Alfred Best, interview, March 8, 2012.

172 parked alongside the wrecked PN9E: Balchen, Ford, and La Farge, *War below Zero*, p. 60.

172 "a hole [to] crawl in.": Spina, memoir, p. 19.

172 dug a "room": Spencer, statement, p. 3.

173 "Glacier Worms": Balchen, Ford, and La Farge, *War below Zero*, p. 62.

173 cut the lines securing the bomber's tail section: *Northland* to SOPA Greenland, December 2, 1942, relayed message from PN9E.

173 "If [supply] plane comes": Ibid., December 1, 1942.

174 losing more weight: Spina, memoir, p. 20.

174 gambling with the navigator's life: Ibid.

174 "In case of emergency, we could travel light": *Northland* to SOPA Greenland, relayed message from PN9E, December 1, 1942.

175 Pritchard, Bottoms, and Howarth: Spina, memoir, p. 20.

175 "Lieutenant O'Hara very ill": Ibid., December 6, 1942. See also Balchen, Ford, and La Farge, *War below Zero*, pp. 66–67.

176 a route recommended by Colonel Balchen: Spencer, statement, p. 3.

176 back within two days: Spina, memoir, p. 20. See also Balchen, Ford, and La Farge, *War below Zero*, p. 66.

176 Tetley said it was time to leave: Spina, memoir, p. 21.

177 Spencer knelt to unstrap his snowshoes: Wedel's fall into the crevasse is described in Balchen, Ford, and La Farge, *War below Zero*, pp. 66–67, and in Spencer, statement, p. 3. See also Lydia McIntosh, "Snowbound for 68 Days." Tetley's account in McIntosh differs slightly, as she quotes him saying they had stopped at night to check on O'Hara. Otherwise the accounts agree.

177 He screamed: Some accounts suggest that Wedel fell without a sound, but O'Hara is quoted as saying he screamed in Francis DeAndrea, "Icy Ordeal Recalled by Crash Survivor," *Scranton Times*, November 9, 1983.

178 marks on a narrow ledge: It seems possible these marks were bloodstains, as some have suggested in various accounts, but La Farge in *The Long Wait* and *War below Zero* never says so directly.

178 He'd never meet his daughter: "McPherson Man Plunged to Death in Greenland Crevasse," *Hutchinson* (Kans.) *News-Herald*, May 27, 1943, scrapbook clipping.

178 "his initiative and perseverance": "Legion of Merit Is Awarded Posthumously to Wichitan," undated newspaper clipping, apparently from the *Kansas City Star*, provided by Eric Langhorst, Wedel's grandson-in-law, March 11, 2012.

179 a gallon of a different grade of lubricating oil: *Northland* to SOPA Greenland, relayed message from PN9E, December 5, 1942.

179 about six miles northeast of the PN9E when the motorsled's engine quit altogether: Balchen, Ford, and La Farge, *War below Zero*, p. 68.

180 tuned to the wrong frequency: Ibid., p. 92.

15: SHOOTING OUT THE LIGHTS

181 "sufficient fuel and supplies for wintering in Comanche Bay": *Northland* to SOPA Greenland, December 8, 1942, message.

181 wanting to stay and needing to go: *Northland* to SOPA Greenland, December 1, 1942.

182 "an overgrown crate, about thirty feet square": "Five Months on the Greenland Ice Cap," *Coast Guard Magazine*, May 1944, p. 28.

182 They spent days tucked in their bunks: Ibid. Also see Taub, *Greenland Ice Cap Rescue*, pp. 5–6. To his credit, Taub spent considerable effort correcting the "official" record, which frequently omitted the efforts of the *Northland* crew members put ashore during the rescue efforts. Although they never reached the PN9E crew or the downed Duck, that should not obscure the hardships they endured in their volunteer effort. It also is worth noting that Fuller's markings on a chart of the Koge Bay area known as H.O. 5773 proved significant for the Duck Hunt. As John Long explained: "Fuller's chart's relevance was not so much for the 2012 mission directly. However,

without it we would not have understood the full dynamics of what took place back in 2008 when we started the Duck Hunt. Consequently, it helped paint the physical picture we see. We were able to put names to geographic locations."

183 all five received commendations: "Five Months on the Greenland Ice Cap," p. 29.

183 "This expedition had to be evaluated": Willoughby, *U.S. Coast Guard*, p. 104.

184 couldn't restart the engine: Spencer, statement, p. 4.

184 ice hole they could use for cooking: Ibid., p. 5.

185 Spencer kept his shovel with him: Balchen, Ford, and La Farge, *War below Zero*, p. 69.

185 a nasty mixture of snow and gasoline: Ibid., p. 70.

185 A bout of diarrhea: Spina, memoir, p. 22.

185 shooting them from the sky: DeAndrea, "Icy Ordeal," p. 3.

186 weren't even certain that all three men were still alive: Ibid., p. 98.

186 couldn't control his dogs: Bernt Balchen, "Operations of Force 4998 A in Connection with PN9E Rescue," memo, April 18, 1943, Corey Ford Papers, Dartmouth.

187 "saw lights moving toward station": Ibid.

187 glider drop-and-snatch scheme: See Mitchell Zuckoff, *Lost in Shangri-La: A True Story of Survival, Adventure, and the Most Incredible Rescue Mission of World War II* (New York: HarperCollins, 2011).

187 "Has Army considered use of auto-gyro or helicopter": Message from CIN-CLANT (Commander-in-Chief of Atlantic Fleet), December 17, 1942.

188 "rejected their use as impracticable": Message from COMINCH (Commander-in-Chief of Atlantic Fleet), December 17, 1942.

188 instructed not to share any details: Krause, interview. Numerous news accounts after the events became public also referred to the secrecy surrounding these events as they happened.

16: SNUBLEBLUSS

194 those two locations are considered the most credible: The small corps of Duck Hunt authorities exchanged hundreds of e-mails on this issue. Agreement was elusive, but Donald Taub deserves credit for placing the "two valid locations" at 65°09' N, 41°01' W and 65°08' N, 41°00' W.

195 "very high degree of certitude": North South Polar presentation to the Coast Guard, initially made in January 2012 and updated multiple times in the following months.

197 "Despite its size and awesome strength": Henning Ting, *Encounters with Wildlife in Greenland* (Nuuk, Greenland: Greenland Home Rule Government Department of Environment and Wildlife Management, 1990), p. 7.

198 "Avoid head shots": Ibid., p. 18.

199 "support for expedition to Greenland": "Solicitation/Contract/Order for Commercial Items," awarded August 13, 2012, order no. HSCGGB-12-P-MAV408.

203 buried in permafrost in 2007: Charles McGrath, "Spirits of the South Pole," *New York Times Sunday Magazine*, July 24, 2011.

17: OUTWITTING THE ARCTIC

204 Canadian bush pilot named Jimmie Wade: Balchen, Ford, and La Farge, *War below Zero*, pp. 74–76. See also Ragnar J. Ragnarsson, *US Navy PBY Catalina Units of the Atlantic War* (Oxford, England: Osprey, 2006), p. 78.

205 Wade received the British Explorer Medal: "Civilians Included in King's Honors," *Montreal Gazette*, June 2, 1943.

206 Five days after Wade and Moe went down: Balchen, "Subject: Search C-53," reads in part: "This officer assigned to temporary duty in command of rescue operations of PN9E."

207 "one last trick to outwit the Arctic": Balchen, *Come North with Me*, p. 242.

207 on rescue missions, it was affectionately called Dumbo: "Battle of the Seas: The Lovely Dumbos," *Time*, August 6, 1945, p. x.

208 "If I'm to crawl in on my hands and knees": Balchen, *Come North with Me*, p. 242.

208 "a glacier-cold shoulder.": Ibid.

209 "desires . . . [PBY] to land on Ice Cap": Message from Admiral Smith, undated but most likely December 31, 1942, or January 1, 1943.

209 "aircraft rescue missions are warranted": Message to COMGREPAT (Commander, Greenland Patrol), January 1, 1943.

209 "[At] no time has it been the intention": Message from COMINCH (Commander in chief of U.S. Navy), January 4, 1943.

210 one more attempt to use a ski-plane: Taub, *Greenland Ice Cap Rescue*, p. 9. Some reports indicate two Beechcraft planes were sent and one disappeared en route from Bluie West One to Bluie East Two.

210 plane's shadow on the ice cap: Herbert Kurz, interview by John Long, April 24, 2009.

210 heating the engine in the frigid predawn hours: Balchen, Ford, and La Farge, *War below Zero*, p. 78.

210 the starter on the number-two engine: Ibid., p. 76.

211 thirty-four supply trips: Matz, *History of the 2nd Ferrying Group*, p. 146.

213 On Christmas Day 1942: Spina, memoir, p. 22a.

213 "We will keep you well supplied": Capt. Kenneth Turner to the PN9E Camp, typewritten note, December 26, 1942. Paul Spina saved the original note and pasted it in his scrapbook.

213 Spencer packed thirty pounds: Matz, *History of the 2nd Ferrying Group*, p. 143.

213 arranging large and small objects in Morse code: Balchen, Ford, and La Farge, *War below Zero*, pp. 70–71.

214 last thing he ever felt: Interview with Patricia O'Hara, August 16, 2012.

214 two bandannas to each stake: Balchen, Ford, and La Farge, *War below Zero*,
    p. 84.

215 an experienced U.S. Army Air Forces dogsled rescue team: Taub, *Greenland
    Ice Cap Rescue*, p. 7. Also, Ragnarsson, *US Navy PBY Catalina Units*, pp. 78–
    79. In his memoir, Balchen spells the pilot's name "Dunlap," but the correct
    spelling is "Dunlop."

215 birthday he shared with his wife: Balchen, Ford, and La Farge, *War below
    Zero*, p. 85.

215 Spencer sprang a birthday surprise: Ibid.

216 "We will try to get you out this time": Ibid., p. 86.

216 "be so near those men": Ibid., p. 79.

217 "For crying out loud": Ibid., p. 86.

217 O'Hara might give up and die: McIntosh, "Snowbound for 68 Days."

19: DUMBO ON ICE

229 Monteverde awoke in the dark: Balchen, Ford, and La Farge, *War below Zero*,
    p. 95. Also, Spina, memoir, p. 24.

230 shook hands and prayed together: Spina, memoir, p. 24.

230 Spina lay awake: Ibid.

230 lifted it from the ice wall: Ibid.

231 they didn't talk about it: Balchen, Ford, and La Farge, *War below Zero*, p. 96.

231 snowed a whopping eighteen feet: Spencer, speech.

231 "Factory indicates forward bulkhead of PBY": Balchen, *Come North with Me*,
    p. 243.

232 "We have had no time to make a test landing": Ibid.

232 Turner radioed down to Harry Spencer on the walkie-talkie: Balchen, Ford,
    and La Farge, *War below Zero*, p. 88.

232 "like a power stall letdown on a glassy sea": Balchen, *Come North with Me*,
    p. 243.

233 strange absence of feeling: Balchen, Ford, and La Farge, *War below Zero*,
    p. 89.

233 "a beautiful sight": Ibid.

233 "light as a bundle of rags": Balchen, *Come North with Me*, p. 244.

234 a specially built stretcher-sled: In his statement (p. 4), Spencer writes, "We
    transported Lt. O'Hara from our quarters to the PBY on a very ingenious
    stretcher ski sled" (p. 4). In Balchen's autobiography, he writes, "I carry him
    to the plane in my arms, as light as a bundle of rags" (p. 244). Balchen's
    statements in his autobiography have come under criticism by Taub and
    others, in part because Balchen describes himself as the pilot on takeoff
    when it is generally agreed that Dunlop flew the plane both ways. However,
    Tetley, in McIntosh, "Snowbound for 68 Days," is quoted as saying, "The
    colonel hit upon a method. By unloading us and letting us push, he could

taxi the plane. He brought it around in circles and kept it moving fast enough to keep from getting stuck." For consistency, this account relies on the reporting of La Farge.

235 After almost two hours of effort: Ragnarsson, *US Navy PBY Catalina Units*, p. 79.

235 ran toward the blister, each one jumping at the last minute: This scene is described by La Farge in Balchen, Ford, and La Farge, *War below Zero*, p. 90; by Balchen in his autobiography, p. 244; and by Tetley in McIntosh, "Snowbound for 68 Days." The only significant difference is who pulls the men inside. In Balchen's account, the blister is manned by Sweetzer and "the radio man," but he does not name him. Tetley also credits "the radio operator" without naming him.

236 "Hello boys, Get on the Walkie-Talkie": Paul Spina saved the original note and pasted it in his scrapbook.

238 all of his fingernails had fallen off: Spina, memoir, p. 24.

238 made them feel warmer: Ibid., p. 25.

238 They seemed to take turns breaking down: Ibid., p. 26.

238 a suicide pact: Ibid.

239 "Why should someone else": Ibid.

239 flew into a rage: Ibid.

239 "a bunch of weaklings": Ibid.

240 "the coldest look I ever seen in my life": Ibid., p. 27.

241 destroy that dream: Ibid.

241 "talking about things drawn from another world": Ibid., p. 28.

21: CROSSED WIRES

257 brains enough to move it: Spina, memoir, p. 28.

258 a hit-and-miss proposition: W. W. Shen, "A History of Antipsychotic Drug Development," *Comprehensive Psychiatry* 40, no. 6 (November–December 1999): 407–14.

258 might have been barbiturates: For a historical discussion of the use of barbiturates as sedatives and anticonvulsants, see Francisco Lopez Munoz, "The History of Barbiturates a Century after Their Clinical Introduction," *Journal of Neuropsychiatric Disease and Treatment* 1, no. 4 (December 2005): 329–43.

259 Monteverde became stuck halfway: Balchen, Ford, and La Farge, *War below Zero*, p. 94.

259 Monteverde was gone awhile: Spina, memoir, p. 29.

259 a dozen roast chickens, pork chops, and cooked steaks: Ibid., p. 30.

260 a natural remedy: Ibid.

260 signs of being delusional: Ibid.

260 the power of prayer: Ibid. Spina recalled the title as "The Power of Prayer," but it almost certainly was "Prayer Is Power," published in *Reader's Digest* in 1941.

260 "It is the only power": From Alexis Carrel, "Prayer Is Power," *Reader's Digest*, March 1941, included in *The Questing Spirit*, by Halford E. Luccock and Frances Brentano (New York: Coward-McCann, 1947).

261 lived on concentrated chocolate bars: Spina, memoir, p. 31.

262 during Admiral Byrd's 1933–1935 Antarctic expedition: Martin Sheridan, "Rescue Chief, from Gloucester, Thinks Trip on Ice Cap Is Fun," *Boston Globe*, May 9, 1944.

262 a square-shaped outcropping of rock named Cape Healey: "Dorchester Picture of the Day," from the Dorchester Atheneum, located at dorchester atheneum.org/page.php?id=3124 (accessed March 19, 2012).

262 a scientific expedition to study penguins: Hendrik Dolleman obituary, www.findagrave.com/cgi-bin/fg.cgi?page=gr&GRid=68397203 (accessed March 19, 2012).

262 graduated from Princeton in 1924: Sheridan, "Rescue Chief."

263 Muscular, tanned, tall, and square-jawed: Hansen, *Greenland's Icy Fury*, p. 19.

263 "an opaque sheet of driving snow particles": Balchen, *Come North with Me*, p. 245.

264 The following day: Balchen, Ford, and La Farge, *War below Zero*, pp. 99–100.

264 climbed atop the front end: Spina, memoir, p. 32.

265 "I guess these ice worms": Ibid., p. 33.

265 how little exhilaration they felt: Ibid.

265 "I guess nothing could excite us": Ibid.

265 They reeked, and they knew it: Ibid., p. 40.

266 throwing it into a crevasse: Ibid.

23: "SOME PLAN IN THIS WORLD"

284 The team's lead dog was Rinsky: Caption to a photo taken by Bernt Balchen, released by the U.S. Army Air Forces Public Relations Office.

285 Most seldom barked: Hansen, *Greenland's Icy Fury*, p. 131–34.

285 Spina was the first to falter: Balchen, Ford, and La Farge, *War below Zero*, p. 101.

285 pursuit of the milk can: Spina, memoir, p. 33.

285 The entrance was a large hole: Spina, memoir, p. 35.

286 the Imperial Hotel: Balchen, Ford, and La Farge, *War below Zero*, p. 102. See also Spina, memoir, p. 35.

286 warming blankets for the men: Alfred "Clint" Best, narrative for his family of his time on the ice, typewritten transcript provided by his son, Robert Best, December 27, 1987, p. 6. (Spina also describes the dogs inside the cave in his memoir, pp. 35–36.)

287 Strong decided to get some exercise: Spina, memoir, p, 36.

287 A bigger worry for Balchen: Balchen, *Come North with Me*, p. 245.

287 Both suffered broken ailerons: Ragnarsson, *US Navy PBY Catalina Units*, p. 79.

287 On April 5, 1943: Balchen, *Come North with Me*, p. 245.

288 promoted to captain: Spina, memoir, p. 36.

290 steel straps from equipment cases: Ragnarsson, *US Navy PBY Catalina Units*, p. 80. Details of the repairs also come from Spina, memoir, p. 37.

290 holed up in the overgrown snow cave: Spina, memoir, p. 37.

290 both engines for takeoff: Ibid., p. 246. Details of the damage to the engine were also taken from Balchen, Ford, and La Farge, *War below Zero*, p. 102; and Spina, memoir, p. 38.

291 The three PN9E survivors were skeptical: Spina, memoir, p. 38.

291 praying for good luck and good weather: Ibid.

291 if its engines failed: Ibid.

292 "If I hadn't flown in this ship before": Ibid.

293 "I have no instruments": Balchen, *Come North with Me*, p. 246.

293 about one thousand feet: Ragnarsson, *US Navy PBY Catalina Units*, p. 80. Spina thought it was more like 600 feet (memoir, p. 39), but Ragnarsson is quoting Dunlop.

294 planning a return to earth: Spina, memoir, p. 39.

294 fifty feet above the ground: Ibid. In Balchen, Ford, and La Farge, *War below Zero*, p. 103, the estimate is fifteen feet.

294 He and Best thought they were about to crash: Best, narrative, p. 3; Spina, memoir, p. 39.

294 patted them on the backs: Spina, memoir, p. 39.

294 far past the danger zone: Ibid. Spina explains that this is based on a conversation between Dunlop and Monteverde, who went to the cockpit during the flight.

295 how much fuel remained: Balchen, Ford, and La Farge, *War below Zero*, p. 103.

295 Larson called Bluie East Two: Spina, memoir, p. 39.

296 prepare for a crash: Ibid. This account is confirmed by Balchen, Ford, and La Farge, *War below Zero*, as well as by Ragnarsson.

296 yet another problem: Ragnarsson, *US Navy PBY Catalina Units*, p. 80.

297 emptied to greet them: Spina, memoir, p. 40.

297 "out to the rescue": Best, narrative, p. 6.

EPILOGUE: AFTER GREENLAND

318 "preferably in the South Pacific": "Fortress Pilot Tells of 148 Days."

318 went to the White House: President Roosevelt, diary, www.fdrlibrary .marist.edu/daybyday (accessed March 14, 2012).

319 this imagined exchange: Paul Peters, *Nine Men against the Arctic*, radio play script, presented on *The Cavalcade of America*, Monday, August 2, 1943.

320 "high devotion to duty": "Greenland Crash Hero Gets Coveted Award," *Los Angeles Times*, September 11, 1943.

321 twenty-two years in the air force: Armand Monteverde obituary, *Daily Republic*, January 9, 1988.

321 "I have not been where I could pay my dues": Harry Spencer to Boy Scouts of America, Dallas Circle Ten Council, August 29, 1943.

322 "I have been without toilet paper": Podraza, interview.

322 "the pristine whiteness of the Ice Cap snow": Harry Spencer, written recollections of his return to the ice cap in June 1989, courtesy of Carol Sue Spencer Podraza.

324 graduated with honors from Georgetown University Law School: "Attorney William O'Hara Dies."

324 resented needing a cane: Patricia O'Hara, interview, August 16, 2012.

324 "The Army has some screwy regulation": "Attorney William O'Hara Dies."

325 "All I have left is the pain and suffering": DeAndrea, "Icy Ordeal," p. 3.

325 "I haven't dwelled on what happened": Ibid.

325 "Anytime it was a bad situation": Jean Spina Gaffney, interview, March 11, 2012.

325 graduated magna cum laude: Alfred Clinton Best obituary, *Houston Chronicle*, March 15, 2002.

326 "They called him 'Kinderpa'": Robert C. Best, interview.

326 refused to fly: Peter Tucciarone, son of Alexander Tucciarone, interview, February 25, 2012.

326 "I would do anything": Kilday, "Survivor Recalls Rescue," *Mobile* (Ala.) *Press*, n.d.

326 "How can I tell them what's in my heart": Ibid.

327 "the circulation still isn't back to normal": Lloyd Puryear to Alexander Tucciarone, April 2, 1943, provided by Peter Tucciarone.

328 "ill with a lung ailment": Pearl Puryear to Angelina Tucciarone, December 15, 1943, provided by Peter Tucciarone.

328 "one of its favorite and most beloved native sons": Lloyd Puryear obituary, *News-Journal*, January 13, 1944.

328 "under rigorous Arctic conditions": Distinguished Flying Cross citation for Lieutenant Bernard W. Dunlop, May 8, 1943.

328 served as a lawyer: Nancy Dunlop, daughter of Bernard Dunlop, interview, March 23, 2012.

328 promoted to major in July 1944: "Military Promotions," *Salt Lake Tribune*, July 25, 1944.

329 went to officer candidate school in Miami: "Glacier Hero Gets Officer's Bars," *San Antonio Light*, September 1, 1943.

329 "Bernt Balchen Saves 7 on Ice Cap": *Chicago Tribune*, May 4, 1943.

329 "Flier of the Snows": "Flier of the Snows," unsigned editorial, *New York Times*, May 5, 1943.

329 secret orders to wipe out a German weather station: Balchen, *Come North with Me*, pp. 246–47; Matz, *History of the 2nd Ferrying Group*, p. 145.

330 flown over the North Pole: Balchen, *Come North with Me*, p. 66. In his 1958 autobiography, Balchen recounts a conversation with Byrd's pilot, Floyd Bennett, who died in 1928. In Balchen's telling, Bennett confirms Balchen's suspicions that their plane, the *Josephine Ford*, wasn't capable of reaching the North Pole.

330 turned back well short of the pole: Raimund E. Goerler, "Richard E. Byrd and the North Pole Flight of 1926: Fact, Fiction as Fact, and Interpretation," monograph, 1999, darchive.mblwhoilibrary.org/bitstream /handle/1912/1918/proc98363.pdf?sequence=1 (accessed April 12, 2012). See also Carroll V. Glines, review of *To the Pole: The Diary and Notebook of Richard E. Byrd, 1925–1927*, by Richard E. Byrd, edited by Raimund E. Goerler, *Virginia Magazine of History and Biography* 107, no. 3 (Summer 1999): 332–33.

330 "a lifetime of remarkable achievements": 106th Congress, 2nd session, Joint Resolution 36, passed October 23, 1999.

330 "the key to the solution of a baffling problem": Francois E. Matthew, obituary for Max Harrison Demorest, *Science*, n.s., 97, no. 2510 (February 5, 1943): 132.

330 "In the death of Max Demorest": Ibid.

331 working for the U.S. Geological Survey: "Reno Woman Takes New Job in Washington," *Reno Evening Gazette*, February 2, 1954.

331 studied botany and geology: Photo caption labeled "Now a Student at the University of Michigan," *Reno Evening Gazette*, November 13, 1957.

331 wrote to the six remaining PN9E survivors: Major James McFarland, Memorial Division, Office of the Quartermaster General, to Alfred C. Best, October 15, 1947. Similar letters were sent to Harry Spencer, Paul Spina, and other crewmen.

331 "Crevasses which we observed": Harry Spencer to the Office of the Quartermaster General, October 25, 1947.

331 had a dream: Reba Greathead, daughter of Clarence Wedel, interview by e-mail, March 14, 2012.

332 "Passed from Earth to Glory": Photograph of the tombstone provided by Eric Langhorst, Wedel's grandson-in-law, March 6, 2012.

332 "War in all its shattering bitterness": "Word Received of L. Howarth Death," *Wausaukee Independent*, February 26, 1943.

332 chased a Nazi vessel: "USCGC *Northland* (WPG–49) History Sketch," Public Affairs Division, U.S. Coast Guard Headquarters, p. 3.

332 purchased by American Zionists: Ibid. See also Ya'acov Friedler, "Aliya Bet Ship Sold for Scrap," news clipping found in U.S. Coast Guard historical files, February 23, 1962, no publication noted.

333 "Our old sister": Ibid.

SELECT BIBLIOGRAPHY

B-17F Bomber Pilot's Flight Operating Instructions. Originally published by the U.S. Army Air Forces, December 1942. Reprinted by Periscopefilm.com.

Balchen, Bernt. *Come North with Me.* New York: E. P. Dutton, 1958.

Balchen, Bernt, Corey Ford, and Oliver La Farge. *War below Zero.* New York: Houghton Mifflin, 1944.

Carlson, William S. *Greenland Lies North.* New York: Macmillan, 1940.

The Coast Guard at War: Greenland Patrol. Washington, D.C.: U.S. Coast Guard, 1945.

Erlich, Gretel. *This Cold Heaven: Seven Seasons in Greenland.* New York: Vintage, 2003.

Hansen, Wallace. *Greenland's Icy Fury.* College Station: Texas A&M University Press, 1994.

Hayes, David. *The Lost Squadron.* Edison, N.J.: Chartwell, 2008.

Howarth, David. *The Sledge Patrol: A WWII Epic of Escape, Survival and Victory.* New York: Macmillan, 1957.

Johnson, Corydon M. *Erection and Maintenance Instructions for Model Grumman J2F-4 Airplanes,* August 16, 1939. U.S. Coast Guard historical archives.

Kearns, David A. *Where Hell Freezes Over: A Story of Amazing Survival and Bravery.* New York: Thomas Dunne, 2005.

Kpomassie, Tete-Michel. *An African in Greenland.* New York: NYRB Classics, 2001.

La Farge, Oliver. *The Eagle in the Egg.* New York: Houghton Mifflin, 1949.

Matz, Onas P. *History of the 2nd Ferrying Group.* Seattle: Modet, 1993.

Novak, Thaddeus D. *Life and Death on the Greenland Patrol, 1942.* Gainesville: University Press of Florida, 2005.

Ostrom, Thomas. *The United States Coast Guard in World War II: A History of Domestic and Overseas Actions.* Jefferson, N.C.: McFarland, 2009.

Seaver, Kirsten. *The Frozen Echo: Greenland and the Exploration of North America, ca. A.D. 1000–1500.* Palo Alto, Calif.: Stanford University Press, 1997.

Taub, Capt. Donald M., USCG Retired. *The Greenland Ice Cap Rescue of B-17 "PN9E," November 5, 1942, to May 8, 1943* (monograph). Washington, D.C.: U.S. Coast Guard History Program, 2011.

Ting, Henning. *Encounters with Wildlife in Greenland.* Nuuk, Greenland: Greenland Home Rule Government Department of Environment and Wildlife Management, n.d.

Vaughan, Norman D. *My Life of Adventure.* Mechanicsburg, Pa.: Stackpole, 1995.

Willoughby, Malcolm F. *U.S. Coast Guard in World War II.* Annapolis, Md.: U.S. Naval Institute Press, 1957.

INDEX

Note: Page numbers in *italics* refer to illustrations.

INDEX   383

Fennie, Joan, 66
Finnish Air Force, 117
Flying Fortress, *see* B-17
frostbite, 63, 74–76, 144
Fuller, Richard, 338
  at Beach Head Station, 182–83
  and Canadian airmen rescue, 89
  Navy and Marine Corps Medal
    awarded to, 183

Gandhi, Mohandas K. (Mahatma), 81
gangrene:
  dry vs. wet, 76
  and loss of limbs, 142
  and O'Hara, 76, 122, 127, 135, 144,
    169
Garr, Bruno, 211
Geological Society of America, 25
George I (navy plane), 94, 101–2
Germany, and World War II, 14–16
Gibson, Charles, 61
"Gibson Girl" transmitter/kite, 61–62,
  65, 66, 109–11
Glacier Girl, 18, 101, 102, 160, 221,
  274
"Glacier Worms," 173
Goodlet, David, 338
  and crash landing, 80
  on *Northland,* 124
  as pilot, 79–80
  rescue of, 83–90, 89, 114, 127
  and survival, 80–84, 89–90
  traveling with dinghy, 82–86
Green, Harold, 182
Greenland, 9–15
  air crashes in, 17–18, 40, 51, 118,
    153–55, 318
  beauty of, 34
  climate of, 12–13, 45, 63, 124, 187–
    88, 211, 231, 237
  coastline of, 12
  as code "Bluie," 16, 112
  cold in, 63

crash sites (map), *xv*
crevasses in, 57, 67, 69, 70, 73, 111,
  127, 147, 285
December solstice in, 105
and Denmark, 13–14
Encounters with Wildlife, 197–98
fjords of, 35
flying in milk over, 37–39, 48, 50–
  52, 56, 150, 302
foehn (warm wind), 82–83
glaciers on, 11–12, 94, 193, 253, 253
hypothermia in, 28, 88–89, 101, 139,
  185
Ivigtut mine, 14–16, 15
landing and takeoffs on ice cap, 129,
  136–40, 206–7, 232–35, 289–93
Lou's 2010 survey in, 194, 225
map, *xiv*
melting ice cap of, 97–98
military investigations in, 206
military rescues in, 21, 187
military secrets in, 188, 318
permits to search in, 45, 97, 197
pilot challenges in, 119
population of, 11
sastrugi (snow waves) in, 57, 127
settlement of, 9–10, 13, 17
size of, 11
snow buildup in, 94–95, 193, 316
Spencer's return to, 322, 323
survival in, 57, 124
trade with Europe, 10, 13
U.S. air bases in, 16–17, 112, 114,
  118
weather stations for, 14
in World War II, 14–18, 40
Greenland Cooperative Salvage Company, 118
Greenland Patrol, 2, 16, 23, 66, 123,
  126, 130, 151, 332
Grumman Duck, 2, 4–6, 5, 128, 135
  all aboard officially declared dead,
    155, 178

ABOUT THE AUTHOR

MITCHELL ZUCKOFF is the author of *Lost in Shangri-La*, a *New York Times* bestseller and winner of the Winship/PEN New England Award. His previous books include *Robert Altman: The Oral Biography* and *Ponzi's Scheme: The True Story of a Financial Legend*. He has written for national and regional publications and is a former special projects reporter for the *Boston Globe*, where he was a Pulitzer Prize finalist for investigative reporting. He is a professor of journalism at Boston University and lives outside Boston.